JOURNAL OF APPLIED LOGICS - IFCOLOG JOURNAL OF LOGICS AND THEIR APPLICATIONS

Volume 11, Number 2

March 2024

Disclaimer

Statements of fact and opinion in the articles in Journal of Applied Logics - IfCoLog Journal of Logics and their Applications (JALs-FLAP) are those of the respective authors and contributors and not of the JALs-FLAP. Neither College Publications nor the JALs-FLAP make any representation, express or implied, in respect of the accuracy of the material in this journal and cannot accept any legal responsibility or liability for any errors or omissions that may be made. The reader should make his/her own evaluation as to the appropriateness or otherwise of any experimental technique described.

ISBN 978-1-84890-456-9
ISSN (E) 2631-9829
ISSN (P) 2631-9810

College Publications
Scientific Director: Dov Gabbay
Managing Director: Jane Spurr

http://www.collegepublications.co.uk

SCOPE AND SUBMISSIONS

This journal considers submission in all areas of pure and applied logic, including:

pure logical systems
proof theory
constructive logic
categorical logic
modal and temporal logic
model theory
recursion theory
type theory
nominal theory
nonclassical logics
nonmonotonic logic
numerical and uncertainty reasoning
logic and AI
foundations of logic programming
belief change/revision
systems of knowledge and belief
logics and semantics of programming
specification and verification
agent theory
databases

dynamic logic
quantum logic
algebraic logic
logic and cognition
probabilistic logic
logic and networks
neuro-logical systems
complexity
argumentation theory
logic and computation
logic and language
logic engineering
knowledge-based systems
automated reasoning
knowledge representation
logic in hardware and VLSI
natural language
concurrent computation
planning

This journal will also consider papers on the application of logic in other subject areas: philosophy, cognitive science, physics etc. provided they have some formal content.

Submissions should be sent to Jane Spurr (jane@janespurr.net) as a pdf file, preferably compiled in LaTeX using the IFCoLog class file.

CONTENTS

ARTICLES

Intuitionistic Views on Connexive Constrictible Falsity 125
  Satoru Niki

Esemihoops 159
  Min Liu and Hongxing Liu

Formalization of the Telegrapher's Equations using Higher-Order-Logic Theorem Proving 197
  Elif Deniz, Adnan Rashid, Osman Hasan and Sofiène Tahar

A Categorical Equivalence for Tense Pseudocomplemented Distributive Lattice 237
  Gustavo Pelaitay and Maia Starobinsky

Corrigendum:
  Article from V10, number 6, December 2023, pp. 993-102 253
  Abdirahman Alasow and Marek Perkowski

Intuitionistic Views on Connexive Constructible Falsity

Satoru Niki*

Department of Philosophy I, Ruhr University Bochum
Satoru.Niki@rub.de

Abstract

Intuitionistic logicians generally accept that a negation can be understood as an implication to absurdity. An alternative account of constructive negation is to define it in terms of a primitive notion of falsity. This approach was originally suggested by D. Nelson, who called the operator constructible falsity, as complementing certain constructive aspects of negation. For intuitionistic logicians to be able to understand this new notion, however, it is desirable that constructible falsity has a comprehensive relationship with the traditional intuitionistic negation. This point is especially pressing in H. Wansing's framework of connexive constructible falsity, which exhibits unusual behaviours. From this motivation, this paper enquires what kind of interaction between the two operators can be satisfactory in the framework. We focus on a few natural-looking candidates for such an interaction, and evaluate their relative merits through analyses of their formal properties with both proof-theoretic and semantical means. We in particular note that some interactions allow connexive constructible falsity to provide a different solution to the problem of the failure of the constructible falsity property in intuitionistic logic. An emerging perspective in the end is that intuitionistic logicians may have different preferences depending on whether absurdity is to be understood as the falsehood.

Keywords— Basic systems; Connexive logic; Constructible falsity; Contradictory logics; Intuitionistic logic.

The author would like to thank Heinrich Wansing and two anonymous referees for their helpful comments and suggestions. He would also like to acknowledge the participants of the seminar on contradictory logics at Ruhr University Bochum and the Logic Seminar at Tohoku University for constructive discussions.

*This work was supported by the funding from the European Research Council (ERC) under the European Union's Horizon 2020 research and innovation programme, grant agreement ERC-2020-ADG, 101018280, ConLog.

1   Introduction

The notion of *constructible falsity* (to be denoted by $\sim$) was first introduced by D. Nelson [21] as an operator capturing the constructive procedures to falsify conjunctive and universal statements. Intuitionistic negation (to be denoted by $\neg$), on the other hand, does not fully capture these methods. As a result, while $\vdash \sim(A \wedge B)$ implies $\vdash \sim A$ or $\vdash \sim B$ (*constructible falsity property*) in a Nelsonian system, an analogous property does not hold with respect to $\neg$ in intuitionistic logic.

Constructible falsity can thus be seen as a way to improve the account offered by intuitionistic negation. This does not, however, mean that one accepting such a view has to give up intuitionistic negation as an intuitionistically acceptable operator. Indeed, in Nelson's original system **N3**, intuitionistic negation is definable by taking $\neg A := A \rightarrow \sim A$, as noted by A.A. Markov [17]. Alternatively, one may allow the *absurdity* constant $\bot$ inside a system with constructible falsity, and define intuitionistic negation as an implication to absurdity, i.e. $\neg A := A \rightarrow \bot$. The accommodation of $\bot$ to the language seems acceptable in light of Nelson's remark that distinguishing the two proof methods for the negation of a universal statement affords one to distinguish the meaning of $\sim \forall x A$ and $\forall x A \rightarrow \bot$ [21, p.17]. If one is interested in talking about the meaning of $\bot$ (as part of the latter formula), then it seems unproblematic to have it in one's vocabulary. An axiomatisation of **N3** with both negations as primitive is indeed used by N.N. Vorob'ev [31]. He suggests that such a formalisation is more suitable as a model of mathematical thoughts: his point appears to be that *reductio ad absurdum* used a lot in mathematics corresponds conceptually to $\neg A$ but not to $A \rightarrow \sim A$. It is a primitive procedure independent of refutation (corresponding to $\sim$), and the two negations must be treated as primitive, in order to reflect the primitive status of the procedures.

This relationship between the negations change when the paraconsistent variant **N4** (formulated[1] by A. Almukdad and Nelson [1]) of **N3** is considered. Intuitionistic negation is not definable in **N4**, so the choice of whether to include $\bot$ becomes more significant. The version of **N4** with $\bot$ is commonly denoted by $\mathbf{N4}^{\bot}$, and both logics and their extensions are investigated by S.P. Odintsov [23]. The book [15] by N. Kamide and H. Wansing treats the proof theory of both systems and their neighbours.

Consider now a scenario where an *intuitionistic logician* (here we just mean somebody who understands the connectives of intuitionistic logic, without necessarily being an intuitionist or constructivist.) tries to make sense of constructible falsity. In all three (propositional) systems we have mentioned, it is possible to convert each formula to an equivalent formula in which $\sim$ occurs only in front of prime formulas. So in a sense, the understanding of the 'meaning' of $\sim A$ Nelson refers to is reduced to the understanding of $\sim p$ (and $\sim \bot$ if $\bot$ is taken as primitive). How then can an intuitionistic logician grasp the meaning of $\sim p$?

In the case of **N3**, we have $\sim A \rightarrow \neg A$ as a theorem, and so an intuitionistic logician may understand $\sim p$ as a strengthening of $\neg p$. This is evidenced by how Markov [17] calls constructible falsity *strong negation*. For **N4** (and $\mathbf{N4}^{\bot}$), on the other hand, there is no

---

[1]Equivalent systems were already introduced by D. Prawitz [28] and F. von Kutschera [30]: see [32] for more historical details about **N4**.

such or any other constraint that relates $\sim p$ with $\neg p$ or its negand p. It behaves almost like another propositional variable.[2] Therefore it seems an intuitionistic logician would have a harder time understanding the meaning of $\sim p$ in **N4** (and **N4**$^\perp$) than in **N3**.

An analogous question can be raised for other logics with constructible falsity based on (positive) intuitionistic logic. An especially interesting case is that of the system **C** introduced by Wansing [33]. This system is obtained from **N4** by changing the condition under which an implication is falsified. As a result of this change, **C** satisfies the criteria of *connexive logic* [18, 35]: namely, it validates the theses proposed by Aristotle (AT,AT') and Boethius (BT,BT'):

$$\text{AT: } \sim(\sim A \to A) \quad \text{BT: } (A \to B) \to \sim(A \to \sim B)$$
$$\text{AT': } \sim(A \to \sim A) \quad \text{BT': } (A \to \sim B) \to \sim(A \to B)$$

meanwhile invalidating $(A \to B) \to (B \to A)$ which would hold if $\to$ were a biconditional. A further characteristics of **C** is that it has as theorems a pair of certain formulas A and $\sim A$, i.e. it is a *negation inconsistent*, yet non-trivial system.

C shares with **N4** the characteristics that there is no stipulation for $\sim p$. At the same time, the option of extending it with $\sim A \to \neg A$ is not available: it results in a trivial system because of the negation inconsistency. A different way of extending **C** is proposed by H. Omori and Wansing [26] and later explored Kripke-semantically by G.K. Olkhovikov [24] and algebraically by D. Fazio and Odintsov [9]. The extension **C3** is obtained with the addition of $A \vee \sim A$ as an axiom schema. Since $p \vee \sim p$ holds in **C3**, it is arguably easier for intuitionistic logicians to make sense of $\sim p$ in **C3** than in **C**.[3] On the other hand, they may not be too satisfied with the non-constructivity of the system, such as the failure of the disjunction property.

A question we may ask then is whether there is a satisfactory system of connexive constructible falsity which is more understandable and acceptable for intuitionistic logicians.[4] It is desired that such a system (i) gives a certain stipulation for $\sim p$ as in **C3**, but (ii) remains constructive. In this enquiry, we presuppose the existence of $\perp$ in the language, following the lines of justification mentioned above. Thus more precisely, our concern will be with respect to the expansions[5] **C**^{ab} and **C**$_3^{ab}$ of **C** and **C3** with the absurdity constant.

In this paper, we shall study formal properties of a few extensions of **C**^{ab} which form a natural hierarchy between **C**^{ab} and **C**$_3^{ab}$ when seen through sequent rules. Our aim is

---

[2]Here it might be suggested that an intuitionistic logician can understand the meaning of $\sim p$ by an analogy with the behaviour of propositional variables. Encodability of derivations in an **N4**-style system into a two-sorted λ-calculus [34] seems to also support such a view. This can be an answer, but it would not satisfy him if he expected (perhaps mistakenly) to see something 'negative' in the behaviour of $\sim p$ that would justify him to take it as a negation.

[3]It exhibits a property often ascribed to negation, which may well be understood (without endorsement) as a claim of decidability, perhaps by an analogy with classical negation.

[4]As one reviewer pointed out, such a system may be seen to motivate connexive logics from the viewpoint of intuitionistic logic, thus has an affinity with the discussion in [37]. On the other hand, our focus is not directly on the connexive theses themselves.

[5]This type of expansions is already studied in [9], but there is a slight variation, as we shall discuss in the next section.

thereby to find out which notion of connexive constructible falsity is more satisfactory for intuitionistic logicians. We shall concentrate on two candidates for the axiom schemata. The first is the schema of *potential omniscience* $\neg\neg(A \vee \sim A)$, which was introduced and investigated by I. Hasuo and R. Kashima [11] in the context of $\mathbf{N4}^{\perp}$. Also, as pointed out by A. Avellone et al. [2], the constructive logic of classical truth by P. Miglioli et al. [19] can be seen as $\mathbf{N3}$ plus potential omniscience, when the classical truth is identified with the intuitionistic double negation. Within the context of $\mathbf{C}$, this schema already appears in the proof of [9, Theorem 49]. The second candidate is the axiom schema $\neg A \rightarrow \sim A$ whose implication is dual to $\sim A \rightarrow \neg A$; we shall call the schema *weak negation* on this ground.

The structure of this paper is as follows: Section 2 introduces Hilbert-style systems and sequent calculi for $\mathbf{C}^{ab}$ and its extensions, and shows their equivalence. Section 3 treats Kripke semantics for the systems, and establishes the soundness and completeness of the systems with respect to the sequent calculi following the general argument presented by O. Lahav and A. Avron [16]. Section 4 then applies the results so far to observe some properties (with an emphasis on negation inconsistency) of the extensions with potential omniscience/weak negation, which can be informative for the evaluation of the systems by intuitionistic logicians. In Section 5, we introduce another type of sequent calculus, formulated originally for $\mathbf{N4}$ in [15], with better proof-theoretic properties. In particular, we show that the calculus for potential omniscience enjoys the subformula property. In Section 6, we make an observation concerning the relation between $\sim$ and $\neg$, which provides a new perspective on the connexive constructible falsity. Lastly, section 7 sums up the insights to evaluate the relative advantages of the systems.

2 Proof Systems

In this section, we shall introduce Hilbert-style axiomatic systems as well as Gentzen-style sequent calculi for the logics that concern us.

2.1 Hilbert-style Systems

The main language $\mathcal{L}$ we shall consider in this paper is defined by the following form.

$$A ::= p \mid \sim A \mid (A \wedge A) \mid (A \vee A) \mid (A \rightarrow A) \mid \perp.$$

If we remove $\perp$ from the definition, it defines another propositional language $\mathcal{L}^{+}$. In both languages, $(A \leftrightarrow B) := (A \rightarrow B) \wedge (B \rightarrow A)$ and in $\mathcal{L}$, $\neg A := A \rightarrow \perp$. The set of all formulas in $\mathcal{L}$ ($\mathcal{L}^{+}$) will be denoted by Form (Form^{+}). The set of subformulas of a formula A and of a set Γ of formulas will be denoted by $Sub(A)$ and $Sub(\Gamma)$.

The *complexity* $c(A)$ of a formula A is defined by the following clauses: $c(p) = c(\perp) = 0$, $c(\sim A) = c(A) + 1$ and $c(A \circ B) = c(A) + c(B) + 2$ for $\circ \in \{\wedge, \vee, \rightarrow\}$.

We first introduce Wansing's system $\mathbf{C}$ [33] in $\mathcal{L}^{+}$ and its expansion $\mathbf{C}^{ab}$ in $\mathcal{L}$ which becomes the basis of our enquiry.

Definition 2.1. The system $\mathbf{C}$ in $\mathcal{L}^{+}$ is defined by the next axiom schemata and a rule.

$$(A\rightarrow(B\rightarrow C))\rightarrow((A\rightarrow B)\rightarrow(A\rightarrow C)) \quad \text{(S)}$$
$$A\rightarrow(B\rightarrow A) \quad \text{(K)}$$
$$A\rightarrow(B\rightarrow(A\wedge B)) \quad \text{(CI)}$$
$$(A_1\wedge A_2)\rightarrow A_i \quad \text{(CE)}$$
$$A_i\rightarrow(A_1\vee A_2) \quad \text{(DI)}$$
$$(A\rightarrow C)\rightarrow((B\rightarrow C)\rightarrow((A\vee B)\rightarrow C)) \quad \text{(DE)}$$

$$\sim(A\wedge B)\leftrightarrow(\sim A\vee\sim B) \quad \text{(NC)}$$
$$\sim(A\vee B)\leftrightarrow(\sim A\wedge\sim B) \quad \text{(ND)}$$
$$\sim(A\rightarrow B)\leftrightarrow(A\rightarrow\sim B) \quad \text{(NI)}$$
$$\sim\sim A\leftrightarrow A \quad \text{(NN)}$$
$$\frac{A \qquad A\rightarrow B}{B} \quad \text{(MP)}$$

A *derivation* in **C** of A from a set of formulas Γ is a finite sequence $B_1,\ldots,B_n \equiv A$ such that each B_i is either an instance of one of the axiom schemata, an element of Γ, or obtained from the preceding elements by means of (MP). Then for derivability, we write $\Gamma \vdash_h \Delta$ if there is a derivation of a disjunction $A_1 \vee \ldots \vee A_n$ from Γ, where Δ is a non-empty set of formulas and $A_1,\ldots,A_n \in \Delta$.

We shall write $A_1,\ldots,A_m,\Gamma \vdash_h \Delta,B_1,\ldots,B_n$ for $\{A_1,\ldots,A_m\}\cup\Gamma \vdash_h \Delta\cup\{B_1,\ldots,B_n\}$. If Δ is a singleton, we will occasionally omit the parentheses.

The second system is obtained from **C** by expanding the language.

Definition 2.2. The system $\mathbf{C}^{ab}$ in $\mathcal{L}$ is defined from the axiomatisation of **C** by an additional axiom schema:

$$\bot \rightarrow A \qquad\qquad \text{(EFQ)}$$

The relation $\Gamma \vdash_{hab} \Delta$ is defined as before, except that we allow Δ to be empty. $\Gamma \vdash_{hab} \emptyset$ will mean that there is a derivation of $\bot$ from Γ.

We now define a few more systems from $\mathbf{C}^{ab}$. Among these, $\mathbf{C}_3^{ab}$ is an expansion of the system **C3** [25] with $\bot$.

Definition 2.3. The systems $\mathbf{C}_{po}^{ab}$, $\mathbf{C}_{wn}^{ab}$ and $\mathbf{C}_3^{ab}$ are each defined with a respective additional axiom schema.

$$\neg\neg(A \vee \sim A) \qquad \text{(PO)}$$
$$\neg A \rightarrow \sim A \qquad \text{(WN)}$$
$$A \vee \sim A \qquad\qquad \text{(3)}$$

We shall use $\vdash_{hpo}$, $\vdash_{hwn}$ and $\vdash_{h3}$ for the derivability of the systems.

Remark 2.4. One possible option in defining the systems is to have $\sim\bot$ (or equivalently, $A \rightarrow \sim\bot$) as an additional axiom schema, as is done in the case of $\mathbf{N4}^{\bot}$, see e.g. [15, 23, 22]. Intuitively, it states that what is absurd is false. This option is indeed adopted in the systems $\mathbf{C}^{\bot}$, $\mathbf{C3}^{\bot}$ and their extensions in [9]. On the other hand, we are *not* assuming this, chiefly due to $\sim\neg A$ being one of its consequences. This means that every intuitionistic negation is false, which seems to be a very strong claim.[6] For another reason, $\sim\bot$ is actually provable in the current definition of $\mathbf{C}_{wn}^{ab}$ and $\mathbf{C}_3^{ab}$, so for these systems we do not need the formula

---

[6]T.M. Ferguson [10] however suggests that it may be possible to motivate the feature using an adequate Brouwer-Heyting-Kolmogorov interpretation.

as an axiom schema (thus $\mathbf{C}_3^{ab}$ is equivalent to $\mathbf{C3}^\perp$). This suggests that each system has already in mind, so to speak, whether and what to say about the falsity of $\perp$. This may be worthwhile to be respected (we shall have a few more words on this topic in the conclusion).

Since (MP) is the only rule present in the systems, it is straightforward to observe that the deduction theorem holds for each of the above systems.

Theorem 2.5. For $* \in \{ab, po, wn, 3\}$ we have:

$$\Gamma, A \vdash_{h*} B \text{ if and only if } \Gamma \vdash_{h*} A \to B.$$

Proof. The 'only if' direction is shown by induction on the depth of derivation. The 'if' direction follows by (MP). □

It is helpful at this stage to note the (non-strict) relative strength of the systems.

Proposition 2.6. The following statements hold.

(i) If $\Gamma \vdash_{hab} \Delta$ then $\Gamma \vdash_{hpo} \Delta$.

(ii) If $\Gamma \vdash_{hpo} \Delta$ then $\Gamma \vdash_{hwn} \Delta$.

(iii) If $\Gamma \vdash_{hwn} \Delta$ then $\Gamma \vdash_{h3} \Delta$.

Proof. (i) is immediate from the definition; (ii) follows since it follows from (WN) that $\vdash_{hwn} \neg{\sim}A \to \neg\neg A$, from which (PO) follows. For (iii), from (3) it follows that $\vdash_{h3} (A \to {\sim}A) \to {\sim}A$, and also $\vdash_{h3} \neg A \to (A \to {\sim}A)$ by (EFQ); so (WN) follows. □

2.2 Sequent Calculi

We shall next introduce (multi-succedent) sequent calculi for $\mathbf{C}^{ab}$, $\mathbf{C}_{po}^{ab}$ and $\mathbf{C}_{wn}^{ab}$. We shall use the framework in which each *sequent* $\Gamma \Rightarrow \Delta$ is such that Γ and Δ are finite *sets*[7] of formulas (cf. e.g. the system $\mathbf{LJ}^{\{\ \}}$ in [3, p.64]). The empty set will be denoted by a blank.

First we define the calculus for $\mathbf{C}^{ab}$.

Definition 2.7. The calculus $\mathbf{GC}^{ab}$ is defined by the following rules.

$$A \Rightarrow A \text{ (Ax)}$$

$$\perp \Rightarrow \text{ (L}\perp) \qquad\qquad \frac{\Gamma \Rightarrow \Delta, A \qquad A, \Gamma' \Rightarrow \Delta'}{\Gamma, \Gamma' \Rightarrow \Delta, \Delta'} \text{ (Cut)}$$

$$\frac{\Gamma \Rightarrow \Delta}{A, \Gamma \Rightarrow \Delta} \text{ (LW)} \qquad\qquad \frac{\Gamma \Rightarrow \Delta}{\Gamma \Rightarrow \Delta, A} \text{ (RW)}$$

---

[7]In this setting, it is important to note that different sequents can be derived from the same rule when applied to the same sequent. For instance, consider (L${\sim}{\sim}$) applied to $\{A, B\} \Rightarrow \{C\}$: then we can derive $\{{\sim}{\sim}A, B\} \Rightarrow \{C\}$, but we may also derive $\{{\sim}{\sim}A, A, B\} \Rightarrow \{C\}$, if the antecedent set is conceived as $\{A\} \cup \{A, B\}$.

$$\dfrac{A_i, \Gamma \Rightarrow \Delta}{A_1 \wedge A_2, \Gamma \Rightarrow \Delta} \ (\text{L}\wedge) \qquad \dfrac{\Gamma \Rightarrow \Delta, A \qquad \Gamma \Rightarrow \Delta, B}{\Gamma \Rightarrow \Delta, A \wedge B} \ (\text{R}\wedge)$$

$$\dfrac{A, \Gamma \Rightarrow \Delta \qquad B, \Gamma \Rightarrow \Delta}{A \vee B, \Gamma \Rightarrow \Delta} \ (\text{L}\vee) \qquad \dfrac{\Gamma \Rightarrow \Delta, A_i}{\Gamma \Rightarrow \Delta, A_1 \vee A_2} \ (\text{R}\vee)$$

$$\dfrac{\Gamma \Rightarrow \Delta, A \qquad B, \Gamma' \Rightarrow \Delta'}{A \rightarrow B, \Gamma, \Gamma' \Rightarrow \Delta, \Delta'} \ (\text{L}\rightarrow) \qquad \dfrac{A, \Gamma \rightarrow B}{\Gamma \Rightarrow A \rightarrow B} \ (\text{R}\rightarrow)$$

$$\dfrac{{\sim}A, \Gamma \Rightarrow \Delta \qquad {\sim}B, \Gamma \Rightarrow \Delta}{{\sim}(A \wedge B), \Gamma \Rightarrow \Delta} \ (\text{L}{\sim}\wedge) \qquad \dfrac{\Gamma \Rightarrow \Delta, {\sim}A_i}{\Gamma \Rightarrow \Delta, {\sim}(A_1 \wedge A_2)} \ (\text{R}{\sim}\wedge)$$

$$\dfrac{{\sim}A_i, \Gamma \Rightarrow \Delta}{{\sim}(A_1 \vee A_2), \Gamma \Rightarrow \Delta} \ (\text{L}{\sim}\vee) \qquad \dfrac{\Gamma \Rightarrow \Delta, {\sim}A \qquad \Gamma \Rightarrow \Delta, {\sim}B}{\Gamma \Rightarrow \Delta, {\sim}(A \vee B)} \ (\text{R}{\sim}\vee)$$

$$\dfrac{\Gamma \Rightarrow \Delta, A \qquad {\sim}B, \Gamma' \Rightarrow \Delta'}{{\sim}(A \rightarrow B), \Gamma, \Gamma' \Rightarrow \Delta, \Delta'} \ (\text{L}{\sim}\rightarrow) \qquad \dfrac{A, \Gamma \Rightarrow {\sim}B}{\Gamma \Rightarrow {\sim}(A \rightarrow B)} \ (\text{R}{\sim}\rightarrow)$$

$$\dfrac{A, \Gamma \Rightarrow \Delta}{{\sim}{\sim}A, \Gamma \Rightarrow \Delta} \ (\text{L}{\sim}{\sim}) \qquad \dfrac{\Gamma \Rightarrow \Delta, A}{\Gamma \Rightarrow \Delta, {\sim}{\sim}A} \ (\text{R}{\sim}{\sim})$$

where $i \in \{1, 2\}$. We write $\vdash_{gab} \Gamma \Rightarrow \Delta$ if there is a derivation in $\mathbf{GC}^{ab}$ of $\Gamma \Rightarrow \Delta$ from the 0-premise rules, i.e. (Ax), (L$\perp$).

As usual, the formulas in Γ, Δ etc. will be called *contexts*, a non-context formula in the premises of a rule will be called *active*, and a non-context formula in the conclusion of a rule will be called *principal*.

For $\mathbf{C}^{ab}_{po}$ and $\mathbf{C}^{ab}_{wn}$, we have the following calculi.

Definition 2.8. The calculi $\mathbf{GC}^{ab}_{po}$ and $\mathbf{GC}^{ab}_{wn}$ are respectively defined from $\mathbf{GC}^{ab}$ each with an additional rule:

$$\dfrac{A, \Gamma \Rightarrow \qquad {\sim}A, \Gamma \Rightarrow}{\Gamma \Rightarrow} \ (\text{gPO}) \qquad \dfrac{A, \Gamma \Rightarrow \qquad {\sim}A, \Gamma \Rightarrow \Delta}{\Gamma \Rightarrow \Delta} \ (\text{gWN})$$

The relations $\vdash_{gpo}$ and $\vdash_{gwn}$ are defined analogously to $\vdash_{gab}$

A calculus for $\mathbf{C}^{ab}_3$ can also be defined, as is done in [26] for $\mathbf{C3}$, by allowing the succedent of both premises in (gWN) to be non-empty. Hence (PO), (WN) and (3) give a natural hierarchy of sequent rules, which can motivate our focus on the axioms.

We proceed to establish the correspondence between the Hilbert-style systems and the sequent calculi.

Proposition 2.9. Let $* \in \{ab, po, wn\}$ and Γ, Δ be finite sets of formulas. Then $\Gamma \vdash_{h*} \Delta$ if and only if $\vdash_{g*} \Gamma \Rightarrow \Delta$.

Proof. The 'only if' direction is shown by induction on the depth of derivation in the Hilbert-style systems. Here we look at the case of (PO) in $\mathbf{GC}^{ab}_{po}$ and (WN) in $\mathbf{GC}^{ab}_{wn}$. For (PO):

$$\dfrac{\dfrac{\dfrac{A \Rightarrow A}{A \Rightarrow A \lor \sim A} \ (\text{R}\lor)}{A, \neg(A\lor\sim A) \Rightarrow} \ (\text{L}{\to}) \qquad \dfrac{\dfrac{\sim A \Rightarrow \sim A}{\sim A \Rightarrow A \lor \sim A} \ (\text{R}\lor)}{\sim A, \neg(A\lor\sim A) \Rightarrow} \ (\text{L}{\to})}{\dfrac{\neg(A\lor\sim A) \Rightarrow}{\Rightarrow \neg\neg(A\lor\sim A)}} \text{(gPO)}$$
$$\ \ (\text{RW}),(\text{R}{\to})$$

(A double line indicates multiple applications of rules). For (WN), we have:

$$\dfrac{\dfrac{A \Rightarrow A \qquad \bot \Rightarrow}{A, \neg A \Rightarrow} \ (\text{L}{\to}) \qquad \dfrac{\sim A \Rightarrow \sim A}{\sim A, \neg A \Rightarrow \sim A} \ (\text{LW})}{\dfrac{\neg A \Rightarrow \sim A}{\Rightarrow \neg A \to \sim A} \ (\text{R}{\to})} \text{(gWN)}$$

For the 'if' direction, we show by induction on the depth of derivation in the sequent calculi. Here we check the cases of (gPO) for $\mathbf{GC}_{po}^{ab}$ and of (gWN) for $\mathbf{GC}_{wn}^{ab}$. For the former, by I.H. we have $A, \Gamma \vdash_{hpo} \bot$ and $\sim A, \Gamma \vdash_{hpo} \bot$. By Theorem 2.5, $\Gamma \vdash_{hpo} \neg A$ and $\Gamma \vdash_{hpo} \neg \sim A$. Hence $\Gamma \vdash_{hpo} \neg(A \lor \sim A)$ and so by (PO) we conclude $\Gamma \vdash_{hpo} \bot$. For the latter, by I.H. $A, \Gamma \vdash_{hwn} \bot$ and $\sim A, \Gamma \vdash_{hwn} B_1 \lor \ldots \lor B_n$ for some $B_1, \ldots, B_n \in \Delta$. By Theorem 2.5 we obtain $\Gamma \vdash_{hwn} \neg A$ and $\Gamma \vdash_{hwn} \sim A \to (B_1 \lor \ldots \lor B_n)$. Thus by (WN) we conclude $\Gamma \vdash_{hwn} B_1 \lor \ldots \lor B_n$, i.e. $\Gamma \vdash_{hwn} \Delta$. □

3 Semantics

In this subsection, we shall introduce[8] Kripke semantics for $\mathbf{C}^{ab}$, $\mathbf{C}_{po}^{ab}$ and $\mathbf{C}_{wn}^{ab}$, and then show that the systems are sound and complete with the semantics.

3.1 Kripke Semantics

We first introduce a Kripke semantics for $\mathbf{C}^{ab}$. The presentation here is a combination of bilateral-style sequent calculi used for $\mathbf{C}$ in [33] and non-deterministic sequent calculi due to O. Lahav and A. Avron [16]. We have a few more words about the latter in Section 3.3.

Definition 3.1. A $\mathbf{C}^{ab}$-*frame* $\mathcal{F}$ is a pair $(W, \leq)$, where W is a non-empty set and $\leq$ is a pre-ordering on W. A $\mathbf{C}^{ab}$-*model* $\mathcal{M}$ is a pair $(\mathcal{F}, \mathcal{V})$, where $\mathcal{F}$ is a $\mathbf{C}^{ab}$-frame, $\mathcal{V} = \{\mathcal{V}^+, \mathcal{V}^-\}$ where $\mathcal{V}^* : \mathsf{Form} \to \mathcal{P}(W)$ for $* \in \{+, -\}$. We shall write $w \in \mathcal{V}^*(A)$ also as $\mathcal{M}, w \Vdash_{ab}^* A$ ($\mathcal{M}$ will be omitted when it is contextually clear). $\mathcal{V}$ must satisfy a general condition below:

(Upward Closure): $w \Vdash_{ab}^* A$ and $w \leq w'$ implies $w' \Vdash_{ab}^* A$.

for both $* \in \{+, -\}$. Moreover, the next conditions for the connectives must also be satisfied.

---

[8]A semantics for $\mathbf{C}_3^{ab}$ can be similarly given following [26], but it is outside our focus here.

$w \Vdash_{ab}^{+} \bot \Leftrightarrow$ never.

$w \Vdash_{ab}^{+} A \wedge B \Leftrightarrow w \Vdash_{ab}^{+} A$ and $w \Vdash_{ab}^{+} B$. $\qquad w \Vdash_{ab}^{-} A \wedge B \Leftrightarrow w \Vdash_{ab}^{-} A$ or $w \Vdash_{ab}^{-} B$.

$w \Vdash_{ab}^{+} A \vee B \Leftrightarrow w \Vdash_{ab}^{+} A$ or $w \Vdash_{ab}^{+} B$. $\qquad w \Vdash_{ab}^{-} A \vee B \Leftrightarrow w \Vdash_{ab}^{-} A$ and $w \Vdash_{ab}^{-} B$.

$w \Vdash_{ab}^{+} A \rightarrow B \Leftrightarrow \forall w' \geq w(w' \Vdash_{ab}^{+} A \Rightarrow w' \Vdash_{ab}^{+} B).$ $\quad w \Vdash_{ab}^{-} A \rightarrow B \Leftrightarrow \forall w' \geq w(w' \Vdash_{ab}^{+} A \Rightarrow w' \Vdash_{ab}^{-} B).$

$w \Vdash_{ab}^{+} {\sim} A \Leftrightarrow w \Vdash_{ab}^{-} A.$ $\qquad\qquad\qquad w \Vdash_{ab}^{-} {\sim} A \Leftrightarrow w \Vdash_{ab}^{+} A.$

With respect to a $\mathbf{C}^{ab}$-model $\mathcal{M}$ and a sequent $\Gamma \Rightarrow \Delta$, we shall write $\mathcal{M}, w \vDash_{ab} \Gamma \Rightarrow \Delta$ if $\mathcal{M}, w \Vdash_{ab}^{+} A$ for all $A \in \Gamma$ implies $\mathcal{M}, w \Vdash_{ab}^{+} B$ for some $B \in \Delta$. If $\mathcal{M}, w \vDash_{ab} \Gamma \Rightarrow \Delta$ for all w in $\mathcal{M}$, then we shall write $\mathcal{M} \vDash_{ab} \Gamma \Rightarrow \Delta$. Finally, we shall write $\vDash_{ab} \Gamma \Rightarrow \Delta$ if $\mathcal{M} \vDash_{ab} \Gamma \Rightarrow \Delta$ for all $\mathcal{M}$.

Remark 3.2. The forcing relations $\Vdash_{ab}^{+}/\Vdash_{ab}^{-}$ may be seen to represent e.g. the concepts of *verification/falsification* or *(support of) truth/(support of) falsity* [33]. In our scenario, intuitionistic logicians can be assumed to understand the former relation, by identifying it with the forcing relation of intuitionistic Kripke semantics (except the one for $\sim$, which encodes $\Vdash_{ab}^{-}$ in $\Vdash_{ab}^{+}$). The latter relation, on the other hand, needs an explanation, especially when it comes to $\Vdash_{ab}^{-} p$ for which no special restriction is given.

Next, we define Kripke semantics for $\mathbf{C}_{po}^{ab}$ and $\mathbf{C}_{wn}^{ab}$.

Definition 3.3. Kripke semantics for $\mathbf{C}_{po}^{ab}$ and $\mathbf{C}_{wn}^{ab}$ (we shall use the subscripts $_{po}$ and $_{wn}$ for $\Vdash$ and $\vDash$.) are each defined from the one for $\mathbf{C}^{ab}$ by the addition of the following condition of (Potential Omniscience) and (Weak Negation), respectively:

$$(\text{Potential Omniscience}): \forall w' \geq w(w' \nVdash_{po}^{+} A) \text{ implies } \exists x \geq w(x \Vdash_{po}^{-} A).$$

$$(\text{Weak Negation}): \quad \forall w' \geq w(w' \nVdash_{wn}^{+} A) \text{ implies } w \Vdash_{wn}^{-} A.$$

In these semantics, some relationships between e.g. verification and falsification of p are given, so the latter concept should be more easily understood by an intuitionistic logician in terms of the former.

Let us note a difference in character between (Potential Omniscience) and (Weak Negation), despite their similar appearances. The former condition may be restricted to propositional variables and $\bot$, similarly to how upward closure is ensured in ordinary Kripke semantics for intuitionistic logic by requiring it to hold only in the atomic case. In contrast, such a restriction does not generalise for the latter condition.

Proposition 3.4. The following statements hold.

(i) If a $\mathbf{C}^{ab}$-model $\mathcal{M}$ satisfies the following conditions:

$$\forall w' \geq w(w' \nVdash_{ab}^{+} p) \text{ implies } \exists x \geq w(x \Vdash_{ab}^{-} p).$$
$$\forall w' \geq w(w' \nVdash_{ab}^{+} \bot) \text{ implies } \exists x \geq w(x \Vdash_{ab}^{-} \bot).$$

then $\mathcal{M}$ is a $\mathbf{C}_{po}^{ab}$-model.

(ii) There exists a $\mathbf{C}^{ab}$-model which satisfies the following conditions:

$$\forall w' \geq w(w' \nVdash_{ab}^{+} p) \text{ implies } w \Vdash_{ab}^{-} p.$$
$$\forall w' \geq w(w' \nVdash_{ab}^{-} p) \text{ implies } w \Vdash_{ab}^{+} p.$$
$$\forall w' \geq w(w' \nVdash_{ab}^{+} \bot) \text{ implies } w \Vdash_{ab}^{-} \bot.$$
$$\forall w' \geq w(w' \nVdash_{ab}^{-} \bot) \text{ implies } w \Vdash_{ab}^{+} \bot.$$

while not being a $\mathbf{C}_{wn}^{ab}$-model.

Proof. For (i), we show by induction on the complexity of formulas that (Potential Omniscience) is satisfied in $\mathcal{M}$. The cases when $A \equiv p, \bot$ follow from the assumption.

When $A \equiv B \wedge C$, we show the contrapositive. If $\neg \exists x \geq w(x \Vdash_{ab}^{-} B \wedge C)$, it must be the case that $\forall x \geq w(x \nVdash_{ab}^{-} B$ and $x \nVdash_{ab}^{-} C)$ $(*)$ and so $\forall x \geq w(x \nVdash_{ab}^{-} B)$. Also, as one of the I.H., $\forall w' \geq w(w' \nVdash_{ab}^{+} B)$ implies $\exists x \geq w(x \Vdash_{ab}^{-} B)$. Hence we deduce from these that $\exists w' \geq w(w' \Vdash_{ab}^{+} B)$. Fix one such w'. By $(*)$, $\forall y \geq w'(y \nVdash_{ab}^{-} C)$. Then as another I.H., it holds that $\forall w'' \geq w'(w'' \nVdash_{ab}^{+} C)$ implies $\exists y \geq w'(y \Vdash_{ab}^{-} C)$. So $\exists w'' \geq w'(w'' \Vdash_{ab}^{+} C)$. By (Upward Closure), we have $w'' \Vdash^{+} B$ for such w'' as well. Therefore $\neg \forall w' \geq w(w' \nVdash_{ab}^{+} B \wedge C)$, as desired.

When $A \equiv B \vee C$, if $\forall w' \geq w(w' \nVdash_{ab}^{+} B \vee C)$ we have $\forall w' \geq w(w' \nVdash_{ab}^{+} B)$. Thus by one of the I.H. $\exists x \geq w(x \Vdash_{ab}^{-} B)$. Take such an x. Then $\forall x' \geq x(x' \nVdash_{ab}^{+} C)$ and so $\exists y \geq x(y \Vdash_{ab}^{-} C)$ from the other I.H.. By (Upward Closure), $y \Vdash_{ab}^{-} B$ as well; so $\exists x \geq w(x \Vdash_{ab}^{-} B \vee C)$.

When $A \equiv B \rightarrow C$, if $\neg \forall w' \geq w(w' \nVdash_{ab}^{+} C)$ then $w' \Vdash_{ab}^{+} C$ and so $w' \Vdash_{ab}^{+} B \rightarrow C$ for some $w' \geq w$. Hence $\neg \forall w' \geq w(w' \nVdash_{ab}^{+} B \rightarrow C)$; consequently $\forall w' \geq w(w' \nVdash_{ab}^{+} B \rightarrow C)$ implies $\exists x \geq w(x \Vdash_{ab}^{-} B \rightarrow C)$. Otherwise, $\forall w' \geq w(w' \nVdash_{ab}^{+} C)$ and by the I.H $\exists x \geq w(x \Vdash_{ab}^{-} C)$. Hence $\exists x \geq w(x \Vdash_{ab}^{-} B \rightarrow C)$ and therefore $\forall w' \geq w(w' \nVdash_{ab}^{+} B \rightarrow C)$ implies $\exists x \geq w(x \Vdash_{ab}^{-} B \rightarrow C)$ in this case as well.

Finally, when $A \equiv \sim B$, by the I.H. $\forall w' \geq w(w' \nVdash_{ab}^{+} B)$ implies $\exists x \geq w(x \Vdash_{ab}^{-} B)$, contraposing which we obtain $\forall w' \geq w(w' \nVdash_{ab}^{-} B)$ implies $\exists x \geq w(x \Vdash_{ab}^{+} B)$. Therefore $\forall w' \geq w(w' \nVdash_{ab}^{+} \sim B)$ implies $\exists x \geq w(x \Vdash_{ab}^{-} \sim B)$.

For (ii), suppose $\mathcal{M} = ((W, \leq), \mathcal{V})$ is such that $W = \{w, x, y\}$, $\leq$ is the reflexive closure of $\{(w, x), (w, y)\}$, $\mathcal{V}^{+}(p) = \{x\}$, $\mathcal{V}^{-}(p) = \{y\}$, $V^{-}(\bot) = W$ and for compound formulas $\mathcal{V}$ is defined in accordance with the equivalences in Definition 3.1: e.g. set $x \in \mathcal{V}^{-}(A \rightarrow B)$ if for all $y \geq x(y \in \mathcal{V}^{+}(A)$ implies $y \in \mathcal{V}^{-}(B))$. Then the equivalences in Definition 3.1 are naturally satisfied, and (Upward Closure) may be checked by induction on the complexity of formula. Therefore $\mathcal{M}$ is a $\mathbf{C}^{ab}$-model. In addition, $\mathcal{M}$ satisfies the conditions of the proposition at each world. For instance, for the first condition, $\forall u' \geq u(u' \nVdash_{ab}^{+} p)$ implies $u = y$, but $u \Vdash_{ab}^{-} p$.

Now, it is readily observed that $\forall w' \geq w(w' \nVdash_{ab}^{+} p \wedge \sim p)$, but $w \nVdash_{ab}^{-} p \wedge \sim p$. Therefore (Weak Negation) is not satisfied for all formulas in $\mathcal{M}$. □

3.2 Soundness

In the next two subsections, we shall establish the soundness and completeness of the three sequent calculi $\mathbf{GC}^{ab}$, $\mathbf{GC}_{po}^{ab}$ and $\mathbf{GC}_{wn}^{ab}$ with respect to their Kripke semantics. We shall treat the soundness direction in this subsection.

Theorem 3.5 (soundness). Let $* \in \{ab, po, wn\}$. Then $\vdash_{g*} \Gamma \Rightarrow \Delta$ implies $\vDash_* \Gamma \Rightarrow \Delta$.

Proof. For $\mathbf{GC}^{ab}$, we can establish the statement by induction on the depth of derivation. For instance, if the last step in the derivation is an instance of (R$\sim\to$):

$$\frac{A, \Gamma \Rightarrow \sim B}{\Gamma \Rightarrow \sim(A \to B)}$$

then by the I.H. $\mathcal{M} \vDash_{ab} A, \Gamma \Rightarrow \sim B$. Suppose $\mathcal{M}, w \Vdash^+_{ab} C$ for all $C \in \Gamma$ and $\mathcal{M}, w' \Vdash^+_{ab} A$ for $w' \geq w$. Then by (Upward Closure) $\mathcal{M}, w' \Vdash^+_{ab} C$ for all $C \in \{A\} \cup \Gamma$; thus $\mathcal{M}, w' \Vdash^+_{ab} \sim B$ and so $\mathcal{M}, w' \Vdash^-_{ab} B$. Therefore $\mathcal{M}, w \Vdash^-_{ab} A \to B$ and consequently $\mathcal{M}, w \Vdash^+_{ab} \sim(A \to B)$.

For $\mathbf{GC}^{ab}_{po}$, we in addition need to check the case for (gPO):

$$\frac{A, \Gamma \Rightarrow \qquad \sim A, \Gamma \Rightarrow}{\Gamma \Rightarrow}$$

If $\mathcal{M}, w \Vdash^+_{po} B$ for all $B \in \Gamma$, then $\mathcal{M}, w' \Vdash^+_{po} A$ for $w' \geq w$ leads to a contradiction by (Upward Closure) and the I.H.. Hence $\forall w' \geq w(\mathcal{M}, w' \nVdash^+_{po} A)$. But then $\exists w' \geq w(\mathcal{M}, w' \Vdash^-_{po} A)$ by (Potential Omniscience), which again contradicts the I.H.. Therefore $\mathcal{M}, w \vDash_{po} \Gamma \Rightarrow$ for all w, as required.

For $\mathbf{GC}^{ab}_{wn}$, we need to check the case for (gWN).

$$\frac{A, \Gamma \Rightarrow \qquad \sim A, \Gamma \Rightarrow \Delta}{\Gamma \Rightarrow \Delta}$$

If $\mathcal{M}, w \Vdash^+_{wn} B$ for all $B \in \Gamma$, then we infer $\forall w' \geq w(\mathcal{M}, w' \nVdash^+_{wn} A)$ as in the previous case. By (Weak Negation), $\mathcal{M}, w \Vdash^-_{wn} A$; so by the I.H. $\mathcal{M}, w \Vdash^+_{wn} C$ for some $C \in \Delta$. Therefore $\mathcal{M}, w \vDash_{wn} \Gamma \Rightarrow \Delta$. $\square$

We can also connect the Hilbert-style systems and the Kripke semantics.

Corollary 3.6. Let $* \in \{ab, po, wn\}$ and Γ, Δ be finite sets of formulas. Then $\Gamma \vdash_{h*} \Delta$ implies $\vDash_* \Gamma \Rightarrow \Delta$.

Proof. An immediate consequence of Proposition 2.9 and Theorem 3.5. $\square$

3.3 Completeness

The proof of completeness follows the one given by Lahav and Avron [16] for *basic systems*, which are sequent calculi that satisfy a few natural criteria. In [16] the authors present a general framework for formulating a sound and strongly complete Kripke semantics for a calculus in the class, to which $\mathbf{GC}^{ab}$, $\mathbf{GC}^{ab}_{po}$ and $\mathbf{GC}^{ab}_{wn}$ also belong. The argument here is only slightly altered from the outline given in [16], in order to fit the bilateral-style semantical setting.

Let us first introduce some preliminary notions. In what follows, we keep using the abbreviation with $* \in \{ab, po, wn\}$.

Definition 3.7 (maximal set). A *maximal* set (for $\mathbf{GC}^{ab}/\mathbf{GC}^{ab}_{po}/\mathbf{GC}^{ab}_{wn}$) is a pair (Γ, Δ) of sets of formulas, where:

(i) For any finite $\Gamma' \subseteq \Gamma$ and $\Delta' \subseteq \Delta$, $\nvdash_{g*} \Gamma' \Rightarrow \Delta'$.

(ii) If $A \notin \Gamma$ then $\vdash_{g*} A, \Gamma' \Rightarrow \Delta'$ for some finite $\Gamma' \subseteq \Gamma$ and $\Delta' \subseteq \Delta$.

(iii) If $A \notin \Delta$ then $\vdash_{g*} \Gamma' \Rightarrow \Delta', A$ for some finite $\Gamma' \subseteq \Gamma$ and $\Delta' \subseteq \Delta$.

Lemma 3.8. If $\nvdash_{g*} \Gamma' \Rightarrow \Delta'$ for any finite $\Gamma' \subseteq \Gamma$ and $\Delta' \subseteq \Delta$, then there is a maximal set (Γ^m, Δ^m) (for $\mathbf{GC}^{ab}/\mathbf{GC}^{ab}_{po}/\mathbf{GC}^{ab}_{wn}$) such that $\Gamma \subseteq \Gamma^m$ and $\Delta \subseteq \Delta^m$.

Proof. Let $(B_i)_{i \in \mathbb{N}}$ and $(C_i)_{i \in \mathbb{N}}$ be the sets of formulas not occurring in Γ and Δ, respectively. Let $(A_i)_{i \in \mathbb{N}}$ be such that $A_0 := B_0$, $A_1 := C_0$, $A_2 := B_1$, $A_3 := C_1$, ... We define pairs $(\Gamma_i, \Delta_i)_{i \in \mathbb{N}}$ inductively by the following clauses:

$$(\Gamma_0, \Delta_0) := (\Gamma, \Delta).$$

$$(\Gamma_{2i+1}, \Delta_{2i+1}) := \begin{cases} (\Gamma_{2i} \cup \{A_{2i}\}, \Delta_{2i}) & \text{if } \nvdash_{g*} \Gamma' \Rightarrow \Delta' \text{ for any finite} \\ & \quad \Gamma' \subseteq \Gamma_{2i} \cup \{A_{2i}\} \ \& \ \Delta' \subseteq \Delta_{2i}. \\ (\Gamma_{2i}, \Delta_{2i}) & \text{otherwise.} \end{cases}$$

$$(\Gamma_{2i+2}, \Delta_{2i+2}) := \begin{cases} (\Gamma_{2i+1}, \Delta_{2i+1} \cup \{A_{2i+1}\}) & \text{if } \nvdash_{g*} \Gamma' \Rightarrow \Delta' \text{ for any finite} \\ & \quad \Gamma' \subseteq \Gamma_{2i+1} \ \& \ \Delta' \subseteq \Delta_{2i+1} \cup \{A_{2i+1}\}. \\ (\Gamma_{2i+1}, \Delta_{2i+1}) & \text{otherwise.} \end{cases}$$

Let $(\Gamma^m, \Delta^m) := (\bigcup_i \Gamma_i, \bigcup_i \Delta_i)$. We need to check that (i)–(iii) of Definition 3.7 hold for the pair (Γ^m, Δ^m).

For (i), if $\vdash_{g*} \Gamma' \Rightarrow \Delta'$ for some finite $\Gamma' \subseteq \Gamma^m$ and $\Delta' \subseteq \Delta^m$, then there is i such that $\Gamma' \subseteq \Gamma_i$ and $\Delta' \subseteq \Delta_i$. However we can check by induction that this cannot be the case for any i.

For (ii), if $A \notin \Gamma^m$ then $\Gamma \subseteq \Gamma^m$ implies $A \equiv A_{2i}$ for some i. If $\nvdash_{g*} \Gamma' \Rightarrow \Delta'$ for all finite $\Gamma' \subseteq \Gamma_{2i} \cup \{A\}$ and $\Delta' \subseteq \Delta_{2i}$, then $A \in \Gamma_{2i+1} \subseteq \Gamma^m$, a contradiction. So there must be finite $\Gamma' \subseteq \Gamma_{2i} \subseteq \Gamma^m$ and $\Delta' \subseteq \Delta_{2i} \subseteq \Delta^m$ such that $\vdash_{g*} A, \Gamma' \Rightarrow \Delta'$. For (iii), the argument is analogous. $\qquad \square$

Next we introduce the notion of a canonical model.

Definition 3.9 (canonical model). The *canonical model* $\mathcal{M}_c = ((W_c, \leq_c), \mathcal{V}_c)$ for $\mathbf{GC}^{ab}$ (or $\mathbf{GC}^{ab}_{po}/\mathbf{GC}^{ab}_{wn}$) is defined by:

- $W_c := \{(\Gamma, \Delta) : (\Gamma, \Delta) \text{ is a maximal set}\}$.

- $(\Gamma, \Delta) \leq_c (\Gamma', \Delta')$ iff $\Gamma \subseteq \Gamma'$.

- $(\Gamma, \Delta) \in \mathcal{V}_c^+(A)$ iff $A \in \Gamma$ and $(\Gamma, \Delta) \in \mathcal{V}_c^-(A)$ iff $\sim A \in \Gamma$.

Towards completeness, we shall show a few lemmas.

Lemma 3.10 (properties of canonical model). Let $\mathcal{M}_c$ be the canonical model for $\mathbf{GC}^{ab}$ (or $\mathbf{GC}_{po}^{ab}/\mathbf{GC}_{wn}^{ab}$). Then:

(i) The following are equivalent.

    (a) $\mathcal{M}_c, (\Gamma, \Delta) \vDash_* \Sigma \Rightarrow \Pi$.

    (b) $\Sigma \nsubseteq \Gamma$ or $\Pi \nsubseteq \Delta$.

    (c) There are finite $\Gamma' \subseteq \Gamma$ and $\Delta' \subseteq \Delta$ such that $\vdash_{g*} \Gamma', \Sigma \Rightarrow \Pi, \Delta'$.

(ii) If $\mathcal{M}_c, (\Gamma', \Delta') \vDash_* \Sigma \Rightarrow \Pi$ for all $(\Gamma', \Delta') \geq_c (\Gamma, \Delta)$, then there is a finite $\Gamma'' \subseteq \Gamma$ such that $\vdash_{g*} \Gamma'', \Sigma \Rightarrow \Pi$.

(iii) $\vdash_{g*} \Sigma \Rightarrow \Pi$ iff $\mathcal{M}_c \vDash_* \Sigma \Rightarrow \Pi$.

Proof. For (i), we shall first check that (a) holds if and only if (b) holds. From (a) to (b), suppose $(\Gamma, \Delta) \vDash_* \Sigma \Rightarrow \Pi$, i.e. $(\Gamma, \Delta) \Vdash_*^+ A$ for all $A \in \Sigma$ implies $(\Gamma, \Delta) \Vdash_*^+ B$ for some $B \in \Pi$. From the definition of $\mathcal{V}_c$, this can be rephrased as that $\Sigma \subseteq \Gamma$ implies $\Gamma \cap \Pi \neq \emptyset$. Hence $\Sigma \subseteq \Gamma$ and $\Pi \subseteq \Delta$ implies $\Gamma \cap \Delta \neq \emptyset$, which contradicts the maximality of (Γ, Δ). Therefore $\Sigma \nsubseteq \Gamma$ or $\Pi \nsubseteq \Delta$. From (b) to (a), if $\Sigma \nsubseteq \Gamma$ then $(\Gamma, \Delta) \nVdash_*^+ A$ for some $A \in \Sigma$. So $(\Gamma, \Delta) \vDash_* \Sigma \Rightarrow \Pi$. If on the other hand $\Pi \nsubseteq \Delta$, then there is $A \in \Pi$ such that $A \notin \Delta$. If in addition $A \notin \Gamma$, then by the definition of a maximal set it must be that $\vdash_{g*} A, \Gamma_1 \Rightarrow \Delta_1$ and $\vdash_{g*} \Gamma_2 \Rightarrow \Delta_2, A$ for some finite $\Gamma_1, \Gamma_2 \subseteq \Gamma$ and $\Delta_1, \Delta_2 \subseteq \Delta$. Hence by (Cut) $\vdash_{g*} \Gamma_1, \Gamma_2 \Rightarrow \Delta_1, \Delta_2$; but this contradicts the maximality of (Γ, Δ). So $A \in \Gamma$ and consequently $(\Gamma, \Delta) \vDash_* \Sigma \Rightarrow \Pi$ as well.

Next we check that (b) holds if and only if (c) holds. From (b) to (c), suppose $\Sigma \nsubseteq \Gamma$ or $\Pi \nsubseteq \Delta$. Consider the former case. Then there is $A \in \Sigma$ such that $A \notin \Gamma$. Now because (Γ, Δ) is maximal, it must be that $\vdash_{g*} A, \Gamma' \Rightarrow \Delta'$ for some $\Gamma' \subseteq \Gamma$ and $\Delta' \subseteq \Delta$. Hence $\vdash_{g*} \Gamma', \Sigma \Rightarrow \Pi, \Delta'$ with respect to the Γ' and Δ'. The latter case is analogous. From (c) to (b), if $\Sigma \subseteq \Gamma$ and $\Pi \subseteq \Delta$, then $\vdash_{g*} \Gamma', \Sigma \Rightarrow \Pi, \Delta'$ contradicts the maximality of (Γ, Δ). Hence $\Sigma \nsubseteq \Gamma$ or $\Pi \nsubseteq \Delta$.

For (ii), we show the contrapositive. Suppose for all $\Gamma' \subseteq \Gamma$ we have $\nvdash_{g*} \Gamma', \Sigma \Rightarrow \Pi$. Then for any $\Sigma' \subseteq \Gamma \cup \Sigma$ and $\Pi' \subseteq \Pi$ it holds that $\nvdash_{g*} \Sigma' \Rightarrow \Pi'$. Hence apply Lemma 3.8 to obtain a maximal set (Σ'', Π'') such that $\Gamma \cup \Sigma \subseteq \Sigma''$ and $\Pi \subseteq \Pi''$. Now by (i), $(\Sigma'', \Pi'') \nvDash_* \Sigma \Rightarrow \Pi$ and $(\Gamma, \Delta) \leq_c (\Sigma'', \Pi'')$.

For (iii), if $\vdash_{g*} \Sigma \Rightarrow \Pi$ then by (i) $(\Gamma, \Delta) \vDash_* \Sigma \Rightarrow \Pi$ for all $(\Gamma, \Delta) \in W_c$. For the converse direction, we show the contrapositive. If $\nvdash_{g*} \Sigma \Rightarrow \Pi$ then apply Lemma 3.8 to obtain a maximal set (Σ', Π'). Then by (i) we conclude $(\Sigma', \Pi') \nvDash_* \Sigma \Rightarrow \Pi$. $\square$

Lemma 3.11. The canonical model for $\mathbf{GC}^{ab}$ is indeed a $\mathbf{C}^{ab}$-model.

Proof. It is readily checked that $(W_c, \leq_c)$ is a non-empty pre-ordered set. For (Upward Closure), if $(\Gamma, \Delta) \Vdash_{ab}^* A$ and $(\Gamma', \Delta') \geq_c (\Gamma, \Delta)$ then $A \in \Gamma \subseteq \Gamma'$ for $* = +$ and $\sim A \in \Gamma \subseteq \Gamma'$ for $* = -$. So $(\Gamma', \Delta') \Vdash_{ab}^* A$.

We also need to check the conditions on $\perp$ and compound formulas.

137

$\bot$ If we have $\bot \in \Gamma$ for some $(\Gamma, \Delta) \in W_c$, then the fact that $\vdash_{gab} \bot \Rightarrow \Delta'$ for any finite $\Delta' \subseteq \Delta$ contradicts the maximality of (Γ, Δ). Hence $\bot \notin \Gamma$ and consequently $(\Gamma, \Delta) \not\Vdash^+_{ab} \bot$ for all $(\Gamma, \Delta) \in W_c$.

$\sim$ For $\Vdash^+_{ab}$, it holds that $(\Gamma, \Delta) \Vdash^+_{ab} \sim A$ iff $\sim A \in \Gamma$ iff $(\Gamma, \Delta) \Vdash^-_{ab} A$. For $\Vdash^-_{ab}$, if $(\Gamma, \Delta) \Vdash^-_{ab} \sim A$ but $(\Gamma, \Delta) \not\Vdash^+_{ab} A$ then $(\Gamma, \Delta) \vDash_{ab} A \Rightarrow$. By Lemma 3.10 (i) there are $\Gamma' \subseteq \Gamma$ and $\Delta' \subseteq \Delta$ such that $\vdash_{gab} \Gamma', A \Rightarrow \Delta'$. Thus by $(L\sim\sim) \vdash_{gab} \Gamma', \sim\sim A \Rightarrow \Delta'$. Hence by Lemma 3.10 (i) again, $(\Gamma, \Delta) \vDash_{ab} \sim\sim A \Rightarrow$. But we have $(\Gamma, \Delta) \Vdash^+_{ab} \sim\sim A$ as $\sim\sim A \in \Gamma$, a contradiction. Therefore $(\Gamma, \Delta) \Vdash^+_{ab} A$. Conversely, if $(\Gamma, \Delta) \Vdash^+_{ab} A$ then $(\Gamma, \Delta) \vDash_{ab} \Rightarrow A$. By Lemma 3.10 (i) there are $\Gamma' \subseteq \Gamma$ and $\Delta' \subseteq \Delta$ such that $\vdash_{gab} \Gamma' \Rightarrow A, \Delta'$. Apply $(R\sim\sim)$ to obtain $\vdash_{gab} \Gamma' \Rightarrow \sim\sim A, \Delta'$. By Lemma 3.10 (i), $(\Gamma, \Delta) \vDash_{ab} \Rightarrow \sim\sim A$. Therefore $(\Gamma, \Delta) \Vdash^-_{ab} \sim A$.

$\wedge$ For $\Vdash^+_{ab}$, suppose $(\Gamma, \Delta) \Vdash^+_{ab} A \wedge B$ but $(\Gamma, \Delta) \not\Vdash^+_{ab} A$. Then $(\Gamma, \Delta) \vDash_{ab} A \Rightarrow$. By Lemma 3.10 (i) there are $\Gamma' \subseteq \Gamma$ and $\Delta' \subseteq \Delta$ such that $\vdash_{gab} \Gamma', A \Rightarrow \Delta'$. Thus by $(L\wedge)$, $\vdash_{gab} \Gamma', A \wedge B \Rightarrow \Delta'$. Hence by Lemma 3.10 (i) again, $(\Gamma, \Delta) \vDash_{ab} A \wedge B \Rightarrow$, a contradiction. Thus $(\Gamma, \Delta) \Vdash^+_{ab} A$ and similarly $(\Gamma, \Delta) \Vdash^+_{ab} B$. Conversely, if $(\Gamma, \Delta) \Vdash^+_{ab} A$ and $(\Gamma, \Delta) \Vdash^+_{ab} B$, then $(\Gamma, \Delta) \vDash_{ab} \Rightarrow A$ and $(\Gamma, \Delta) \vDash_{ab} \Rightarrow B$. By Lemma 3.10 (i), there are $\Gamma' \subseteq \Gamma$ and $\Delta' \subseteq \Delta$ such that $\vdash_{gab} \Gamma' \Rightarrow A, \Delta'$ and $\vdash_{gab} \Gamma' \Rightarrow B, \Delta'$. Thus by $(R\wedge)$, $\vdash_{gab} \Gamma' \Rightarrow A \wedge B, \Delta'$. So by Lemma 3.10 (i), $(\Gamma, \Delta) \vDash_{ab} \Rightarrow A \wedge B$. Therefore $(\Gamma, \Delta) \Vdash^+_{ab} A \wedge B$.

Next for $\Vdash^-_{ab}$, if $(\Gamma, \Delta) \Vdash^-_{ab} A \wedge B$ then $(\Gamma, \Delta) \not\Vdash^-_{ab} A$ and $(\Gamma, \Delta) \not\Vdash^-_{ab} B$ imply $(\Gamma, \Delta) \vDash_{ab} \sim A \Rightarrow$ and $(\Gamma, \Delta) \vDash_{ab} \sim B \Rightarrow$. By Lemma 3.10 (i) there are $\Gamma' \subseteq \Gamma$ and $\Delta' \subseteq \Delta$ such that $\vdash_{gab} \Gamma', \sim A \Rightarrow \Delta'$ and $\vdash_{gab} \Gamma', \sim B \Rightarrow \Delta'$. By $(L\sim\wedge)$ we infer $\vdash_{gab} \Gamma', \sim(A \wedge B) \Rightarrow \Delta'$; so by Lemma 3.10 (i) again, $(\Gamma, \Delta) \vDash_{ab} \sim(A \wedge B) \Rightarrow$. Hence $\sim(A \wedge B) \notin \Gamma$ and so $(\Gamma, \Delta) \not\Vdash^-_{ab} A \wedge B$, a contradiction. Therefore either $(\Gamma, \Delta) \Vdash^-_{ab} A$ or $(\Gamma, \Delta) \Vdash^-_{ab} B$. Conversely, if $(\Gamma, \Delta) \Vdash^-_{ab} A$ then $\sim A \in \Gamma$ and so $(\Gamma, \Delta) \vDash_{ab} \Rightarrow \sim A$. Then $(\Gamma, \Delta) \vDash_{ab} \Rightarrow \sim(A \wedge B)$ by Lemma 3.10 (i) and $(R\sim\wedge)$. Therefore $(\Gamma, \Delta) \Vdash^-_{ab} A \wedge B$. The case when $(\Gamma, \Delta) \Vdash^-_{ab} B$ is analogous.

$\vee$ Similar to the cases for conjunction.

$\rightarrow$ For $\Vdash^+_{ab}$, suppose $(\Gamma, \Delta) \Vdash^+_{ab} A \rightarrow B$. Then since $\vdash_{gab} A, A \rightarrow B \Rightarrow B$, by Lemma 3.10 (i) we infer $(\Gamma', \Delta') \vDash_{ab} A \Rightarrow B$ for any $(\Gamma', \Delta') \geq_c (\Gamma, \Delta)$; that is to say, $(\Gamma', \Delta') \Vdash^+_{ab} A$ implies $(\Gamma', \Delta') \Vdash^+_{ab} B$ for all $(\Gamma', \Delta') \geq_c (\Gamma, \Delta)$.

Conversely, if for all $(\Sigma, \Pi) \geq_c (\Gamma, \Delta)$ it holds that $(\Sigma, \Pi) \Vdash^+_{ab} A$ implies $(\Sigma, \Pi) \Vdash^+_{ab} B$, then by Lemma 3.10 (ii) we infer $\vdash_{gab} \Gamma', A \Rightarrow B$ for some $\Gamma' \subseteq \Gamma$. By $(R\rightarrow)$ we obtain $\vdash_{gab} \Gamma' \Rightarrow A \rightarrow B$. Hence by Lemma 3.10 (i) we conclude $(\Gamma, \Delta) \vDash_{ab} \Rightarrow A \rightarrow B$, i.e. $(\Gamma, \Delta) \Vdash^+_{ab} A \rightarrow B$. The case for $\Vdash^-_{ab}$ is argued in a similar manner, using the already established equivalence for negation. $\square$

Lemma 3.12. The canonical model for $\mathbf{GC}^{ab}_{po}$ ($\mathbf{GC}^{ab}_{wn}$) is indeed a $\mathbf{C}^{ab}_{po}$-model ($\mathbf{C}^{ab}_{wn}$-model).

Proof. For $\mathbf{GC}^{ab}_{po}$, we have to check that the canonical model satisfies (Potential Omniscience). Towards a contradiction, suppose that $(\Gamma', \Delta') \not\Vdash^+_{po} A$ for all $(\Gamma', \Delta') \geq_c (\Gamma, \Delta)$ but $(\Gamma', \Delta') \not\Vdash^-_{po} A$ for all $(\Gamma', \Delta') \geq_c (\Gamma, \Delta)$. Then $(\Gamma', \Delta') \vDash_{po} A \Rightarrow$ and $(\Gamma', \Delta') \vDash_{po} \sim A \Rightarrow$ for each such (Γ', Δ'); hence by Lemma 3.10 (ii) we conclude $\vdash_{gpo} \Sigma, A \Rightarrow$ and $\vdash_{gpo} \Sigma, \sim A \Rightarrow$

for some $\Sigma \subseteq \Gamma$. By (gPO) , $\vdash_{gpo} \Sigma \Rightarrow$; thus $(\Gamma, \Delta) \vDash_{po} \Rightarrow$ by Lemma 3.10 (i), a contradiction. Therefore we can conclude that $(\Gamma', \Delta') \Vdash^-_{po} A$ for some $(\Gamma', \Delta') \geq_c (\Gamma, \Delta)$.

For $\mathbf{GC}^{ab}_{wn}$, we have to check (Weak Negation). Suppose that $(\Gamma', \Delta') \nVdash^+_{wn} A$ for all $(\Gamma', \Delta') \geq_c (\Gamma, \Delta)$. Then like in the previous case, $\vdash_{gwn} \Sigma, A \Rightarrow$ for some $\Sigma \subseteq \Gamma$. Also $\vdash_{gwn} \Sigma, {\sim}A \Rightarrow {\sim}A$. Thus by (gWN) $\vdash_{gwn} \Sigma \Rightarrow {\sim}A$; therefore $(\Gamma, \Delta) \vDash_{wn} \Rightarrow {\sim}A$ and so $(\Gamma, \Delta) \Vdash^-_{wn} A$. $\qquad \square$

Now we are ready to show the completeness theorem.

Theorem 3.13 (completeness). Let $* \in \{ab, po, wn\}$. If $\vDash_* \Gamma \Rightarrow \Delta$ then $\vdash_{g*} \Gamma \Rightarrow \Delta$.

Proof. Suppose $\vDash_* \Gamma \Rightarrow \Delta$. Consider the canonical model $\mathcal{M}_c$, which is by Lemma 3.11 and 3.12 is an appropriate model. Then by Lemma 3.10 (iii), we conclude $\vdash_{g*} \Gamma \Rightarrow \Delta$. $\qquad \square$

We consequently obtain the completeness with respect to Hilbert-style systems as well.

Corollary 3.14. Let $* \in \{ab, po, wn\}$ and Γ, Δ be finite sets of formulas. Then $\vDash_* \Gamma \Rightarrow \Delta$ implies $\Gamma \vdash_{h*} \Delta$.

4 Properties of $\mathbf{C}^{ab}_{po}$ and $\mathbf{C}^{ab}_{wn}$

In this section, we shall look at some properties of $\mathbf{C}^{ab}_{po}$ and $\mathbf{C}^{ab}_{wn}$ that can be shown from the results we have established so far. They provide useful information when we later discuss which of the systems an intuitionistic logician might prefer.

We begin with separating the Hilbert-style systems. This gives a strict hierarchy of the systems, $\mathbf{C}^{ab}_3$, $\mathbf{C}^{ab}_{wn}$, $\mathbf{C}^{ab}_{po}$ and $\mathbf{C}^{ab}$ when ordered from the strongest to the weakest.

Proposition 4.1. The following statements hold.

  (i) $\nvdash_{hab} \neg\neg(A \vee {\sim}A)$.

  (ii) $\nvdash_{hpo} \neg A \rightarrow {\sim}A$.

  (iii) $\nvdash_{hwn} A \vee {\sim}A$.

Proof. For (i), take a model $\mathcal{M} = ((W, \leq), \mathcal{V})$ such that $W = \{w, w'\}$; $\leq$ is the reflexive closure of $\{(w, w')\}$; and $\mathcal{V}$ is defined inductively such that $\mathcal{V}^+(p) = \mathcal{V}^-(p) = \emptyset$ for all p, $\mathcal{V}^-(\bot) = W$, and for compound formulas, $\mathcal{V}^+$ and $\mathcal{V}^-$ are defined according to the equivalences in Definition 3.1. As before, $\mathcal{M}$ is easily checked to be a $\mathbf{C}^{ab}$-model. Now, since $\mathcal{M}, w' \nVdash^+_{ab} p$ and $\mathcal{M}, w' \nVdash^+_{ab} {\sim}p$, it holds that $\mathcal{M}, w' \nVdash^+_{ab} p \vee {\sim}p$. Thus $\mathcal{M} \nvDash_{ab} \Rightarrow \neg\neg(p \vee {\sim}p)$. The statement then holds by Corollary 3.6.

For (ii), consider a model $\mathcal{M}'$ defined analogously to $\mathcal{M}$ with the only difference being that $\mathcal{V}^-(p) = \{w'\}$ for all p. In order to check that $\mathcal{M}'$ is a $\mathbf{C}^{ab}_{po}$-model, we need to show that (Potential Omniscience) holds. First observe that:

$$\mathcal{M}', w' \nVdash^+_{po} p \Rightarrow \mathcal{M}', w' \Vdash^-_{po} p;$$
$$\mathcal{M}', w' \nVdash^-_{po} p \Rightarrow \mathcal{M}', w' \Vdash^+_{po} p.$$

Now if for some $x \in W$ it holds that $\forall x' \geq x(\mathcal{M}', x' \not\Vdash_{po}^{+} p)$, then $\mathcal{M}', w' \not\Vdash_{po}^{+} p$; so $\mathcal{M}', w' \Vdash_{po}^{-} p$. Hence $\exists y \geq x(y \Vdash_{po}^{-} p)$. From Proposition 3.4 (i), this and the easily checkable case for $\bot$ are sufficient to establish (Potential Omniscience).

The only thing that is left is to observe that $\mathcal{M}'$ works as a counter-model. For this it suffices to note that $\mathcal{M}', w' \not\Vdash_{po}^{+} p$ implies $\mathcal{M}', w \Vdash_{po}^{+} \neg p$, but $\mathcal{M}', w \not\Vdash_{po}^{+} \sim p$; Therefore $\mathcal{M}', w \not\Vdash_{po}^{+} \neg p \to \sim p$.

For (iii), consider a model $\mathcal{M}''$ defined analogously to $\mathcal{M}'$ with the only difference being that $\mathcal{V}^{+}(p) = \{w'\}$ as well, for all p. In order to check that $\mathcal{M}''$ is a $\mathbf{C}_{wn}^{ab}$-model, we claim that

$$\mathcal{M}'', w' \not\Vdash_{wn}^{+} A \Rightarrow \mathcal{M}'', w \Vdash_{wn}^{-} A;$$
$$\mathcal{M}'', w' \not\Vdash_{wn}^{-} A \Rightarrow \mathcal{M}'', w \Vdash_{wn}^{+} A.$$

The cases when $A \equiv p, \bot$ are immediate. For conjunction, if $w' \not\Vdash_{wn}^{+} B \wedge C$ then $w' \not\Vdash_{wn}^{+} B$ or $w' \not\Vdash_{wn}^{+} C$. Hence by the I.H. $w \Vdash_{wn}^{-} B$ or $w \Vdash_{wn}^{-} C$ and so $w \Vdash_{wn}^{-} B \wedge C$. For the second item, if $w' \not\Vdash_{wn}^{-} B \wedge C$ then $w' \not\Vdash_{wn}^{-} B$ and $w' \not\Vdash_{wn}^{-} C$. By the I.H. $w \Vdash_{wn}^{+} B$ and $w \Vdash_{wn}^{+} C$; so $w \Vdash_{wn}^{+} B \wedge C$. The cases for disjunction are analogous.

For implication, if $w' \not\Vdash_{wn}^{+} B \to C$ then $w' \not\Vdash_{wn}^{+} C$. By the I.H. $w \Vdash_{wn}^{-} C$ and so $w \Vdash_{wn}^{-} B \to C$. The case for the second item is similar. Finally if $w' \not\Vdash_{wn}^{+} \sim A$ then $w' \not\Vdash_{wn}^{-} A$. By the I.H. $w \Vdash_{wn}^{+} A$ and so $w \Vdash_{wn}^{-} \sim A$. The case for the second item is similar as well.

Now if for some x it holds that $\forall x' \geq x(\mathcal{M}'', x' \not\Vdash_{wn}^{+} A)$, then $\mathcal{M}'', w' \not\Vdash_{wn}^{+} A$ and so by the claim $\mathcal{M}'', w \Vdash_{wn}^{-} A$, which implies $\mathcal{M}'', x \Vdash_{wn}^{-} A$, as required. Finally, to see that $\mathcal{M}''$ invalidates $A \vee \sim A$, note that $\mathcal{M}'', w \not\Vdash^{+} p \vee \sim p$. $\qquad\square$

Another point that follows from soundness is that $\sim\bot$ is not a theorem of $\mathbf{C}_{po}^{ab}$.

Proposition 4.2. $\vdash_{hpo} \neg\neg\sim\bot$ but $\nvdash_{hpo} \sim\bot$.

Proof. The former follows from $\sim\bot \leftrightarrow (\bot \vee \sim\bot)$ and $\neg\neg(\bot \vee \sim\bot)$. For the latter, construct a model $\mathcal{M} = ((W, \leq), \mathcal{V})$ where $W = \{w, w'\}$, $\leq$ is the reflexive closure of $\{(w, w')\}$, $\mathcal{V}^{+}(p) = \mathcal{V}^{-}(p) = W$, $\mathcal{V}^{-}(\bot) = \{w'\}$ and otherwise $\mathcal{V}$ is defined according to the equivalences in Definition 3.1. Then it is straightforward to see via Proposition 3.4 that $\mathcal{M}$ is a $\mathbf{C}_{po}^{ab}$-model. Note then $\mathcal{M}, w \not\Vdash_{po}^{-} \bot$. Hence $\nvdash_{hpo} \sim\bot$ by Corollary 3.6. $\qquad\square$

Remark 4.3. This observation also implies that $\vdash_{hpo} \neg\neg\sim\neg A$ but $\nvdash_{hpo} \sim\neg A$. Since the latter may be a controversial formula, it can be taken as an advantage that $\mathbf{C}_{po}^{ab}$ does not prove it. The provability of the former, on the other hand, appears less controversial, because it merely states that the falsity of an intuitionistic negation does not (in the sense of $\neg$) lead to absurdity. It might be even preferable from the perspective of an intuitionistic logician, because it offers some information about the status of $\sim\bot$ compared with the case for $\mathbf{C}^{ab}$ where it is left unspecified.

Next we shall point to the strength of $\mathbf{C}_{po}^{ab}$ in expressing *provable contradictions*, i.e. formulas A such that both A and $\sim A$ are provable in the system. For this purpose, we shall use the logic **CN** introduced by J. Cantwell [4] plus $\perp$, an expansion already considered in [9]. As pointed out by Omori and Wansing [25], **CN** can be seen as an extension of **C3** with Peirce's law; for more information about **CN** and related systems, see also the two-part papers by P. Égré, L. Rossi and J. Sprenger [6, 7] as well as [8].

Definition 4.4. We define $\mathbf{CN}^{\perp}$ by adding the next axiom schema to $\mathbf{C}_3^{ab}$:

$$((A \to B) \to A) \to A \tag{PL}$$

We shall use $\vdash_{hcn}$ to denote the derivability.

$\mathbf{CN}^{\perp}$ works as the classical counterpart of $\mathbf{C}_{po}^{ab}$, as confirmed by expanding Glivenko's theorem to include $\sim$.

Proposition 4.5 (Glivenko's theorem). $\Gamma \vdash_{hcn} A$ if and only if $\Gamma \vdash_{hpo} \neg\neg A$.

Proof. The 'if' direction is immediate. For the 'only if' direction, it follows by induction on the depth of derivation in $\mathbf{CN}^{\perp}$. Most of the cases are as in Glivenko's theorem for intuitionistic logic (see e.g. [27]). The only important case is that of (3), but clearly, (PO) suffices in this case. $\qquad\square$

Using this, we can embed provable contradictions of $\mathbf{CN}^{\perp}$ into $\mathbf{C}_{po}^{ab}$ in a simple manner.

Corollary 4.6. A formulas A is a provable contradiction in $\mathbf{CN}^{\perp}$ if and only if $\neg(A \wedge \sim A) \to A$ is so in $\mathbf{C}_{po}^{ab}$.

Proof. For the 'only if' direction, if A is a provable contradiction in $\mathbf{CN}^{\perp}$ then by Proposition 4.5 we infer $\vdash_{hpo} \neg\neg(A \wedge \sim A)$. hence $\vdash_{hpo} \neg(A \wedge \neg A) \to A$ and $\vdash_{hpo} \neg(A \wedge \neg A) \to \sim A$, so by (NI) $\vdash_{hpo} \sim(\neg(A \wedge \neg A) \to A)$ as well. For the 'if' direction, if $\neg(A \wedge \sim A) \to A$ is a provable contradiction in $\mathbf{C}_{po}^{ab}$ then $\vdash_{hpo} \neg(A \wedge \sim A) \to (A \wedge \sim A)$ and so $\vdash_{hcn} A \wedge \sim A$. Thus the statement follows. $\qquad\square$

Remark 4.7. We may also note that the same embedding does not work with respect to $\mathbf{C}^{ab}$. It is straightforward from (3) that $(p \leftrightarrow \sim p) \to p$ is a provable contradiction in $\mathbf{CN}^{\perp}$, but we can construct a $\mathbf{C}^{ab}$-model $\mathcal{M} = ((\{w\}, (\{(w, w)\})), \mathcal{V})$ such that $\mathcal{V}^+(p) = \mathcal{V}^-(p) = \emptyset$: we can show in this model that $\mathcal{M}, w \nVdash_{ab}^+ \neg(((p \leftrightarrow \sim p) \to p) \wedge \sim((p \leftrightarrow \sim p) \to p)) \to ((p \leftrightarrow \sim p) \to p)$.

The above corollary says that $\mathbf{C}_{po}^{ab}$ is as rich as $\mathbf{CN}^{\perp}$ in producing provable contradictions. As we shall see later, $\mathbf{C}_{po}^{ab}$ is a constructive system, so this means that every provable contradiction in $\mathbf{CN}^{\perp}$ has a constructive counterpart. At the same time, one may wonder whether this is due to (PO) being a rather strong principle. There can be a worry that the system is not acceptable to someone who is interested in provable contradictions but is inclined to stay in **C**. The following observation, based on the conservativity of Jankov's logic over positive intuitionistic logic [13], addresses such worry to some extent.

Proposition 4.8. Let Γ and Δ be finite sets of formulas in $\mathcal{L}^+$. Then $\Gamma \vdash_{hab} \Delta$ iff $\Gamma \vdash_{hpo} \Delta$ iff $\Gamma \vdash_{hwn} \Delta$.

Proof. By Proposition 2.6, it is sufficient to check that $\Gamma \vdash_{hwn} \Delta$ implies $\Gamma \vdash_{hab} \Delta$. We shall show the contrapositive of this implication. If $\Gamma \nvdash_{hab} \Delta$ then $\nvDash_{ab} \Gamma \Rightarrow \Delta$ by Corollary 3.14. Thus there is a $\mathbf{C}^{ab}$-model $\mathcal{M} = ((W, \leq), \mathcal{V})$ such that $\mathcal{M}, w_0 \nvDash_{ab} \Gamma \Rightarrow \Delta$ for some $w_0 \in W$. Define a new model $\mathcal{M}' = ((W', \leq'), \mathcal{V}')$ as follows.

- $W' := W \cup \{u\}$.

- $\leq' := \leq \cup \{(w, u) : w \in W'\}$.

- $\mathcal{V}'^*$ for $* \in \{+, -\}$ is defined inductively, by:

  - $\mathcal{V}'^*(p) := \mathcal{V}^*(p) \cup \{u\}$.
  - $\mathcal{V}'^-(\bot) := W'$.
  - otherwise, the equivalences in Definition 3.1 are followed.

We claim $\mathcal{M}'$ is a $\mathbf{C}^{ab}_{wn}$-model. It is immediate from the definition that all the equivalences in Definition 3.1 hold. Then it is also straightforward to check that (Upward Closure) is satisfied.

We need also to check that (Weak Negation) holds. Towards this, we shall first show by induction that for any $w \in W'$:

$$\mathcal{M}', u \nVdash^+_{wn} A \Longrightarrow \mathcal{M}', w \Vdash^-_{wn} A \text{ and } \mathcal{M}', u \nVdash^-_{wn} A \Longrightarrow \mathcal{M}', w \Vdash^+_{wn} A.$$

Since $\mathcal{M}', u \Vdash^+_{wn} p$ and $\mathcal{M}', u \Vdash^-_{wn} p$ for all p, the cases for propositional variables hold. For $\bot$, the statements hold because $\mathcal{M}', w \Vdash^-_{wn} \bot$ for all $w \in W'$.

For conjunction, first if $\mathcal{M}', u \nVdash^+_{wn} A \wedge B$, then $\mathcal{M}', u \nVdash^+_{wn} A$ or $\mathcal{M}', u \nVdash^+_{wn} B$. By the I.H. $\mathcal{M}', w \Vdash^-_{wn} A$ or $\mathcal{M}', w \Vdash^-_{wn} B$; hence $\mathcal{M}', w \Vdash^-_{wn} A \wedge B$. Next, if $\mathcal{M}', u \nVdash^-_{wn} A \wedge B$ then $\mathcal{M}', u \nVdash^-_{wn} A$ and $\mathcal{M}', u \nVdash^-_{wn} B$. By the I.H. $\mathcal{M}', w \Vdash^+_{wn} A$ and $\mathcal{M}', w \Vdash^+_{wn} B$; hence $\mathcal{M}', w \Vdash^+_{wn} A \wedge B$. The cases for disjunction are analogous.

For implication, first if $\mathcal{M}', u \nVdash^+_{wn} A \to B$ then $\mathcal{M}', u \nVdash^+_{wn} B$. Thus by the I.H. $\mathcal{M}', w \Vdash^-_{wn} B$ and consequently $\mathcal{M}', w \Vdash^-_{wn} A \to B$. Similarly, if $\mathcal{M}', u \nVdash^-_{wn} A \to B$ then $\mathcal{M}', u \nVdash^-_{wn} B$, so by the I.H. $\mathcal{M}', w \Vdash^+_{wn} B$ and $\mathcal{M}', w \Vdash^+_{wn} A \to B$.

For negation, first if $\mathcal{M}', u \nVdash^+_{wn} {\sim} A$ then $\mathcal{M}', u \nVdash^-_{wn} A$. By the I.H. $\mathcal{M}', w \Vdash^+_{wn} A$; hence $\mathcal{M}', w \Vdash^-_{wn} {\sim} A$. Similarly, if $\mathcal{M}', u \nVdash^-_{wn} {\sim} A$ then $\mathcal{M}', u \nVdash^+_{wn} A$. By the I.H. $\mathcal{M}', w \Vdash^-_{wn} A$ and so $\mathcal{M}', w \Vdash^-_{wn} {\sim} A$.

Now, if $\forall w' \geq w(\mathcal{M}', w' \nVdash^+_{wn} A)$ then in particular $\mathcal{M}, u \nVdash^+_{wn} A$. By what we have established, we infer $\mathcal{M}, x \Vdash^-_{wn} A$ for any $x \in W'$. Therefore $\mathcal{M}, w \Vdash^-_{wn} A$. It is thus established that (Weak Negation) is satisfied. Consequently $\mathcal{M}'$ is an $\mathbf{C}^{ab}_{wn}$-model.

In order to establish the proposition itself, we shall observe that

$$\mathcal{M}, w \Vdash^*_{ab} A \text{ if and only if } \mathcal{M}', w \Vdash^*_{wn} A$$

for $* \in \{+, -\}$, $w \in W$ and A in $\mathcal{L}^+$. The cases for propositional variables hold by stipulation.

For conjunction, first, $\mathcal{M}, w \Vdash^+_{ab} A \wedge B$ holds if and only if $\mathcal{M}, w \Vdash^+_{ab} A$ and $\mathcal{M}, w \Vdash^+_{ab} B$. By the I.H. this is equivalent to $\mathcal{M}', w \Vdash^+_{wn} A$ and $\mathcal{M}', w \Vdash^+_{wn} B$ and hence to $\mathcal{M}', w \Vdash^+_{wn} A \wedge B$. Similarly, $\mathcal{M}, w \Vdash^-_{ab} A \wedge B$ holds if and only if $\mathcal{M}, w \Vdash^-_{ab} A$ or $\mathcal{M}, w \Vdash^-_{ab} B$ holds. By the I.H. this is equivalent to that $\mathcal{M}', w \Vdash^-_{wn} A$ or $\mathcal{M}', w \Vdash^-_{wn} B$ and hence to $\mathcal{M}', w \Vdash^-_{wn} A \wedge B$. The cases for disjunction are similar.

For implication, first, $\mathcal{M}, w \Vdash^+_{ab} A \to B$ holds if $\forall w' \geq w (\mathcal{M}, w' \Vdash^+_{ab} A \Rightarrow \mathcal{M}, w' \Vdash^+_{ab} B)$. By the I.H. this is equivalent to $\forall w' \geq w (\mathcal{M}', w' \Vdash^+_{wn} A \Rightarrow \mathcal{M}', w' \Vdash^+_{wn} B)$. Furthermore, as is easily checkable, $\mathcal{M}', u \Vdash^*_{wn} C$ for $* \in \{+, -\}$ and C in $\mathcal{L}^+$. Therefore $\mathcal{M}', u \Vdash^+_{wn} A$ implies $\mathcal{M}', u \Vdash^+_{wn} B$ as well. Thus $\mathcal{M}, w \Vdash^+_{ab} A \to B$ is equivalent to $\forall w' \geq' w (\mathcal{M}, w' \Vdash^+_{wn} A \Rightarrow \mathcal{M}, w' \Vdash^+_{wn} B)$, i.e. $\mathcal{M}', w \Vdash^+_{wn} A \to B$. Similarly for the case for $\Vdash^-$.

For negation, $\mathcal{M}, w \Vdash^+_{ab} \sim A$ holds if and only if $\mathcal{M}, w \Vdash^-_{ab} A$. By the I.H., this is equivalent to $\mathcal{M}', w \Vdash^-_{wn} A$ and therefore to $\mathcal{M}, w \Vdash^+_{ab} \sim A$. The case for $\Vdash^-$ is analogous.

We are now ready to observe that $\mathcal{M}', w_0 \Vdash^+_{wn} A$ for all $A \in \Gamma$ but $\mathcal{M}', w_0 \nVdash^+_{wn} B$ for all $B \in \Delta$. Therefore $\nvDash_{wn} \Gamma \Rightarrow \Delta$ and by Corollary 3.6 we conclude $\Gamma \nvdash_{hwn} \Delta$. $\square$

5 More on Sequent Calculus

In this section, we shall introduce another type of sequent calculi for $\mathbf{C}^{ab}$, $\mathbf{C}^{ab}_{po}$ and $\mathbf{C}^{ab}_{wn}$ which have certain proof-theoretic advantages. We shall show their cut-eliminability and make a few observations related to constructivity and the subformula property.

5.1 Bilateral-style Sequent Calculi

The calculi we shall consider are based on the subformula calculus $\mathbf{Sn4}$ for $\mathbf{N4}$, introduced by N. Kamide and H. Wansing [14, 15]. As the name suggests, this type of calculi shows a better behaviour with respect to the subformula property than the type of calculi of Definition 2.7. In addition, it has a more bilateral flavour (see e.g. [5, 29, 36]) as well, which might be preferable from certain philosophical perspectives.

In this type of calculus, a sequent (for distinction, we shall call it a *b-sequent*) is of the form $\Gamma | \Delta \Rightarrow^* \Pi$, where Γ, Δ are finite sets of formulas, Π is a set of formulas with at most one element, and $* \in \{+, -\}$. Let us first look at a calculus for $\mathbf{C}^{ab}$.

Definition 5.1. The calculus $\mathbf{SC}^{ab}$ is defined by the following rules:

$$A| \Rightarrow^- A \text{ (Ax}-) \qquad\qquad |A \Rightarrow^+ A \text{ (Ax}+)$$

$$|\bot \Rightarrow^* \text{ (L}\bot+)$$

$$\frac{\Gamma | \Delta \Rightarrow^- A \qquad A, \Gamma' | \Delta' \Rightarrow^* \Pi}{\Gamma, \Gamma' | \Delta, \Delta' \Rightarrow^* \Pi} \text{ (Cut}-) \qquad \frac{\Gamma | \Delta \Rightarrow^+ A \qquad \Gamma' | \Delta', A \Rightarrow^* \Pi}{\Gamma, \Gamma' | \Delta, \Delta' \Rightarrow^* \Pi} \text{ (Cut}+)$$

$$\frac{\Gamma | \Delta \Rightarrow^* \Pi}{A, \Gamma | \Delta \Rightarrow^* \Pi} \text{ (LW}-) \qquad\qquad \frac{\Gamma | \Delta \Rightarrow^*}{\Gamma | \Delta \Rightarrow^- C} \text{ (RW}-)$$

$$\frac{\Gamma|\Delta \Rightarrow^* \Pi}{\Gamma|\Delta, A \Rightarrow^* \Pi} \ (\text{LW}+)$$

$$\frac{\Gamma|\Delta \Rightarrow^*}{\Gamma|\Delta \Rightarrow^+ C} \ (\text{RW}+)$$

$$\frac{A, \Gamma|\Delta \Rightarrow^* \Pi \qquad B, \Gamma|\Delta \Rightarrow^* \Pi}{A \wedge B, \Gamma|\Delta \Rightarrow^* \Pi} \ (\text{L}\wedge-)$$

$$\frac{\Gamma|\Delta \Rightarrow^- A_i}{\Gamma|\Delta \Rightarrow^- A_1 \wedge A_2} \ (\text{R}\wedge-)$$

$$\frac{\Gamma|\Delta, A_i \Rightarrow^* \Pi}{\Gamma|\Delta, A_1 \wedge A_2 \Rightarrow^* \Pi} \ (\text{L}\wedge+)$$

$$\frac{\Gamma|\Delta \Rightarrow^+ A \qquad \Gamma|\Delta \Rightarrow^+ B}{\Gamma|\Delta \Rightarrow^+ A \wedge B} \ (\text{R}\wedge+)$$

$$\frac{A_i, \Gamma|\Delta \Rightarrow^* \Pi}{A_1 \vee A_2, \Gamma|\Delta \Rightarrow^* \Pi} \ (\text{L}\vee-)$$

$$\frac{\Gamma|\Delta \Rightarrow^- A \qquad \Gamma|\Delta \Rightarrow^- B}{\Gamma|\Delta \Rightarrow^- A \vee B} \ (\text{R}\vee-)$$

$$\frac{\Gamma|\Delta, A \Rightarrow^* \Pi \qquad \Gamma|\Delta, B \Rightarrow^* \Pi}{\Gamma|\Delta, A \vee B \Rightarrow^* \Pi} \ (\text{L}\vee+)$$

$$\frac{\Gamma|\Delta \Rightarrow^+ A_i}{\Gamma|\Delta \Rightarrow^+ A_1 \vee A_2} \ (\text{R}\vee+)$$

$$\frac{\Gamma|\Delta \Rightarrow^+ A \qquad B, \Gamma'|\Delta' \Rightarrow^* \Pi}{A \rightarrow B, \Gamma, \Gamma'|\Delta, \Delta' \Rightarrow^* \Pi} \ (\text{L}\rightarrow-)$$

$$\frac{\Gamma|\Delta, A \Rightarrow^- B}{\Gamma|\Delta \Rightarrow^- A \rightarrow B} \ (\text{R}\rightarrow-)$$

$$\frac{\Gamma|\Delta \Rightarrow^+ A \qquad \Gamma'|\Delta', B \Rightarrow^* \Pi}{\Gamma, \Gamma'|\Delta, \Delta', A \rightarrow B \Rightarrow^* \Pi} \ (\text{L}\rightarrow+)$$

$$\frac{\Gamma|\Delta, A \Rightarrow^+ B}{\Gamma|\Delta \Rightarrow^+ A \rightarrow B} \ (\text{R}\rightarrow+)$$

$$\frac{\Gamma|\Delta, A \Rightarrow^* \Pi}{\sim A, \Gamma|\Delta \Rightarrow^* \Pi} \ (\text{L}\sim-)$$

$$\frac{\Gamma|\Delta \Rightarrow^+ A}{\Gamma|\Delta \Rightarrow^- \sim A} \ (\text{R}\sim-)$$

$$\frac{A, \Gamma|\Delta \Rightarrow^* \Pi}{\Gamma|\Delta, \sim A \Rightarrow^* \Pi} \ (\text{L}\sim+)$$

$$\frac{\Gamma|\Delta \Rightarrow^- A}{\Gamma|\Delta \Rightarrow^+ \sim A} \ (\text{R}\sim+)$$

where $i \in \{1, 2\}$. The derivability in $\mathbf{SC}^{ab}$ will be denoted by $\vdash_{sab}$. If the rules (Cut$-$) and (Cut$+$) are removed from $\mathbf{SC}^{ab}$, it defines the cut-free system $\mathbf{SC}^{ab}$-(Cut), whose derivability is denoted by $\vdash_{sab}^{cf}$.

Next we define the bilateral-style calculi for $\mathbf{C}_{po}^{ab}$ and $\mathbf{C}_{wn}^{ab}$.

Definition 5.2. The calculus $\mathbf{SC}_{po}^{ab}$ is defined from $\mathbf{SC}^{ab}$ by the following rules.

$$\frac{p, \Gamma|\Delta \Rightarrow^* \qquad \Gamma|\Delta, p \Rightarrow^*}{\Gamma|\Delta \Rightarrow^*} \ (\text{sPO}) \qquad \frac{\perp, \Gamma|\Delta \Rightarrow^*}{\Gamma|\Delta \Rightarrow^*} \ (\text{L}\perp-)$$

The calculus $\mathbf{SC}_{wn}^{ab}$ is defined from $\mathbf{SC}^{ab}$ by the following rule.

$$\frac{A, \Gamma|\Delta \Rightarrow^* \Pi \qquad \Gamma|\Delta, A \Rightarrow^*}{\Gamma|\Delta \Rightarrow^* \Pi} \ (\text{sWN})$$

The derivability in $\mathbf{SC}^{ab}$ and $\mathbf{SC}_{wn}^{ab}$ will be denoted by $\vdash_{spo}$ and $\vdash_{swn}$. The derivability in the cut-free systems $\mathbf{SC}_{po}^{ab}$-(Cut) and $\mathbf{SC}_{wn}^{ab}$-(Cut) will be denoted by $\vdash_{spo}^{cf}$ and $\vdash_{swn}^{cf}$.

The rule (sPO) is modelled after the rule (Gem-at) for the systems $\mathbf{G3ip}$ +(Gem-at) in [20] and $\mathbf{G3C3at}$ in [26]. We should also note already that eliminating cut in $\mathbf{SC}_{wn}^{ab}$ does not give too many benefits, for (gWN) can similarly remove an arbitrary formula.

Before moving onto the proof of cut-elimination, let us observe the correspondence between the bilateral-style calculi and Hilbert-style calculi. For this purpose, we shall use the notations $\sim\Gamma := \{\sim A : A \in \Gamma\}$, $\emptyset^+, \emptyset^- := \bot$, $\{C\}^+ := C$ and $\{C\}^- := \sim C$.

Proposition 5.3. Let $\dagger \in \{ab, po, wn\}$. If $\vdash_{s\dagger} \Gamma|\Delta \Rightarrow^* \Pi$ then $\sim\Gamma, \Delta \vdash_{h\dagger} \Pi^*$.

Proof. By induction on the depth of derivation in $\mathbf{SC}_\dagger^{ab}$. For instance, when the last rule applied is an instance of $(L\to-)$:

$$\frac{\Gamma|\Delta \Rightarrow^+ A \qquad B,\Gamma'|\Delta' \Rightarrow^- C}{A \to B, \Gamma, \Gamma'|\Delta, \Delta' \Rightarrow^- C}$$

Then from the I.H. $\sim\Gamma, \Delta \vdash_{h\dagger} A$ and $\sim B, \sim\Gamma', \Delta' \vdash_{h\dagger} \sim C$. It is now straightforward to observe from (NC) and Theorem 2.5 that $\sim(A \to B), \sim\Gamma, \sim\Gamma', \Delta, \Delta' \vdash_{h\dagger} \sim C$. $\square$

For the other direction, we need a couple of lemmas for $\mathbf{C}_{po}^{ab}$.

Lemma 5.4. The following statements hold.

(i) If $\vdash_{spo} A \wedge B, \Gamma|\Delta \Rightarrow^* C$ then $\vdash_{spo} A, \Gamma|\Delta \Rightarrow^* C$ and $\vdash_{spo} B, \Gamma|\Delta \Rightarrow^* C$.

(ii) If $\vdash_{spo} A \vee B, \Gamma|\Delta \Rightarrow^* C$ then $\vdash_{spo} A, B, \Gamma|\Delta \Rightarrow^* C$.

(iii) If $\vdash_{spo} A \to B, \Gamma|\Delta \Rightarrow^* C$ then $\vdash_{spo} B, \Gamma|\Delta \Rightarrow^* C$.

(iv) If $\vdash_{spo} \sim A, \Gamma|\Delta \Rightarrow^* C$ then $\vdash_{spo} \Gamma|\Delta, A \Rightarrow^* C$.

(v) If $\vdash_{spo} \Gamma|\Delta, A \wedge B \Rightarrow^* C$ then $\vdash_{spo} \Gamma|\Delta, A, B \Rightarrow^* C$.

(vi) If $\vdash_{spo} \Gamma|\Delta, A \vee B \Rightarrow^* C$ then $\vdash_{spo} \Gamma|\Delta, A \Rightarrow^* C$ and $\vdash_{spo} \Gamma|\Delta, B \Rightarrow^* C$.

(vii) If $\vdash_{spo} \Gamma|\Delta, A \to B \Rightarrow^* C$ then $\vdash_{spo} \Gamma|\Delta, B \Rightarrow^* C$.

(viii) If $\vdash_{spo} \Gamma|\Delta, \sim A \Rightarrow^* C$ then $\vdash_{spo} A, \Gamma|\Delta \Rightarrow^* C$.

Proof. By (Cut$-$) and (Cut$+$). $\square$

Lemma 5.5. If $\vdash_{spo} A, \Gamma|\Delta \Rightarrow^*$ and $\vdash_{spo} \Gamma|\Delta, A \Rightarrow^*$ then $\vdash_{spo} \Gamma|\Delta \Rightarrow^*$.

Proof. By induction on the complexity of A. The cases when $A \equiv p$ and $A \equiv \bot$ follow from (sPO) and (L$\bot-$), respectively. If $A \equiv B \wedge C$, then $B \wedge C, \Gamma|\Delta \Rightarrow^*$ and $\Gamma|\Delta, B \wedge C \Rightarrow^*$. By Lemma 5.4 it holds that $B, \Gamma|\Delta \Rightarrow^*$; $C, \Gamma|\Delta \Rightarrow^*$ and $\Gamma|\Delta, B, C \Rightarrow^*$. Hence we obtain the next derivation.

$$\frac{\Gamma|\Delta, B, C \Rightarrow^* \qquad \dfrac{\dfrac{B, \Gamma|\Delta \Rightarrow^*}{B, \Gamma|\Delta, C \Rightarrow^*}\ (\text{LW}+)}{\Gamma|\Delta, C \Rightarrow^*}\ (\text{I.H.}) \qquad C, \Gamma|\Delta \Rightarrow^*}{\Gamma|\Delta \Rightarrow^*}\ (\text{I.H.})$$

The other cases are analogous. $\square$

Proposition 5.6. For $\dagger \in \{ab, po, wn\}$, if $\Gamma \vdash_{h\dagger} A$ then $\vdash_{s\dagger} |\Gamma \Rightarrow^+ A$.

Proof. By induction on the depth of derivation in $\mathbf{C}_{\dagger}^{ab}$. As an example, one direction of (NI) is:

$$
\frac{|A \Rightarrow^+ A \quad \dfrac{\dfrac{B| \Rightarrow^- B}{|\sim B \Rightarrow^- B}\ (\mathrm{L}\sim+)}{|A, A \to \sim B \Rightarrow^- B}\ (\mathrm{L}\to+)}{\dfrac{\dfrac{|A \to \sim B \Rightarrow^- A \to B}{|A \to \sim B \Rightarrow^+ \sim(A \to B)}\ (\mathrm{R}\sim+)}{| \Rightarrow^+ (A \to \sim B) \to \sim(A \to B)}\ (\mathrm{R}\to+)}\ (\mathrm{R}\to-)
$$

The other cases are checked similarly. For (PO), we need to appeal to Lemma 5.5. □

5.2 Cut-elimination

For cut-elimination, the argument will be a standard one, but as in [14, 15], we have to take care of two types of cut rules. We begin with introducing a couple of notions: suppose we have a derivation in $\mathbf{SC}_{\dagger}^{ab}$ ($\dagger \in \{ab, po, wn\}$) in which there is an application of cut (i.e. either (Cut$-$) or (Cut$+$)). Then by the *grade* of the cut, we shall mean the complexity of the cutformula (the formula A in (Cut$-$) and (Cut$+$).) By the *height* of the cut, we shall mean the number of b-sequents that occur in the subderivation which has the conclusion of cut as the endsequent.

Let us first establish a couple of lemmas.

Lemma 5.7. Let $\dagger \in \{ab, po, wn\}$. Then $\vdash_{s\dagger}^{cf} \Gamma|\Delta \Rightarrow^+$ if and only if $\vdash_{s\dagger}^{cf} \Gamma|\Delta \Rightarrow^-$.

Proof. By induction on the depth of derivation. □

Lemma 5.8. Let $\dagger \in \{ab, po, wn\}$ and suppose there is a derivation of a b-sequent in $\mathbf{SC}_{\dagger}^{ab}$ in which (Cut$-$) or (Cut$+$) is applied only at the last step. Then there is a derivation of the b-sequent $\mathbf{SC}_{\dagger}^{ab}$ in which there is no application of (Cut$-$) nor (Cut$+$).

Proof. We shall establish the statement by double induction, with the main induction on the grade of (Cut$-$)/(Cut$+$), and the subinduction on the height of (Cut$-$)/(Cut$+$). We divide into cases depending on which rules are applied to obtain the premises of the (Cut$-$)/(Cut$+$).

First we consider the cases where one of the premises is an instance of one of the 0-premise rules (Ax$-$), (Ax$+$) or (L$\perp$+). Then for the first two cases, the subderivation ending with the other premise is the desired derivation. If the right premise is (L$\perp$+) and the left premise is (RW+), then the subderivation ending with the premise of the (RW+) is either the desired derivation or is different from it only by the sign on the arrow: in this case apply Lemma 5.7. If the right premise is one of the other rules, e.g. (L$\to-$), the derivation must have the following form.

$$\dfrac{\dfrac{\Gamma|\Delta \Rightarrow^+ A \qquad B,\Gamma'|\Delta' \Rightarrow^+ \bot}{A \to B, \Gamma, \Gamma'|\Delta, \Delta' \Rightarrow^+ \bot}\,(L\to-) \qquad |\bot \Rightarrow^*}{A \to B, \Gamma, \Gamma'|\Delta, \Delta' \Rightarrow^*}\,(\text{Cut}+)$$

Then we can construct the following derivation:

$$\dfrac{\Gamma|\Delta \Rightarrow^+ A \qquad \dfrac{B,\Gamma'|\Delta \Rightarrow^+ \bot \qquad |\bot \Rightarrow^*}{B,\Gamma'|\Delta' \Rightarrow^*}\,(\text{Cut}+)}{A \to B, \Gamma, \Gamma'|\Delta, \Delta' \Rightarrow^*}\,(L\to-)$$

Since the new instance of (Cut+) is of lower height, it is possible to apply the I.H. to the subderivation ending with the instance of (Cut+); so we obtain a cut-free derivation of the endsequent.

Secondly, if one of the premises is obtained by an application of a weakening rule (i.e. (LW−), (LW+), (RW−) or (RW+)), then we can argue similarly to the previous cases, along with possible applications of weakening rules.

Thirdly, assume both of the premises are obtained through non-0-premise and non-weakening rules, but the cutformula is not principal in one of them. Consider, as a first example, the case of (Cut−) where the left premise is obtained through (sWN).

$$\dfrac{\dfrac{A,\Gamma|\Delta \Rightarrow^- B \qquad \Gamma|\Delta, A \Rightarrow^-}{\Gamma|\Delta \Rightarrow^- B}\,(\text{sWN}) \qquad B,\Gamma'|\Delta' \Rightarrow^* C}{\Gamma,\Gamma'|\Delta, \Delta' \Rightarrow^* C}\,(\text{Cut}-)$$

Then we can construct the following derivation (the dashed line indicates applications of Lemma 5.7 and weakening).

$$\dfrac{\dfrac{A,\Gamma|\Delta \Rightarrow^- B \qquad B,\Gamma'|\Delta' \Rightarrow^* C}{A, \Gamma, \Gamma'|\Delta, \Delta' \Rightarrow^* C}\,(\text{Cut}-) \qquad \dfrac{\Gamma|\Delta, A \Rightarrow^-}{\Gamma,\Gamma'|\Delta, \Delta', A \Rightarrow^*}}{\Gamma,\Gamma'|\Delta, \Delta' \Rightarrow^* C}\,(\text{sWN})$$

We can then apply the I.H. to the subderivation ending with the instance of (Cut−). As a second example, consider the case of (Cut+) for $\mathbf{SC}^{ab}_{po}$ where the right premise is obtained through (sPO).

$$\dfrac{\Gamma|\Delta \Rightarrow^+ A \qquad \dfrac{p,\Gamma'|\Delta', A \Rightarrow^* \qquad \Gamma'|\Delta', A, p \Rightarrow^*}{\Gamma'|\Delta', A \Rightarrow^*}\,(\text{sPO})}{\Gamma,\Gamma'|\Delta, \Delta' \Rightarrow^*}\,(\text{Cut}+)$$

Then we can construct the next derivation:

$$\dfrac{\dfrac{\Gamma|\Delta \Rightarrow^+ A \qquad p,\Gamma'|\Delta', A \Rightarrow^*}{p,\Gamma,\Gamma'|\Delta, \Delta' \Rightarrow^*}\,(\text{Cut}+) \qquad \dfrac{\Gamma|\Delta \Rightarrow^+ A \qquad \Gamma'|\Delta', A, p \Rightarrow^*}{\Gamma,\Gamma'|\Delta, \Delta', p \Rightarrow^*}\,(\text{Cut}+)}{\Gamma,\Gamma'|\Delta, \Delta' \Rightarrow^*}\,(\text{sPO})$$

Then we can apply the I.H. to the subderivations ending with an application of (Cut+).

Finally, assume that both of the premises are obtained through non-0-premise and non-weakening rules, and the cutformula is principal in both of them. Here we look at the cases for (Cut−) and the cutformula is an implication:

$$\frac{\dfrac{\Gamma|\Delta, A \Rightarrow^- B}{\Gamma|\Delta \Rightarrow^- A \to B}\,(\mathrm{R}{\to}-) \qquad \dfrac{\Gamma'|\Delta' \Rightarrow^+ A \qquad B,\Gamma''|\Delta'' \Rightarrow^* C}{A \to B, \Gamma', \Gamma''|\Delta', \Delta'' \Rightarrow^* C}\,(\mathrm{L}{\to}-)}{\Gamma, \Gamma', \Gamma''|\Delta, \Delta', \Delta'' \Rightarrow^* C}\,(\mathrm{Cut}-)$$

Then we can construct the following derivation:

$$\frac{\dfrac{\Gamma'|\Delta' \Rightarrow^+ A \qquad \Gamma|\Delta, A \Rightarrow^- B}{\Gamma, \Gamma'|\Delta, \Delta' \Rightarrow^- B}\,(\mathrm{Cut}+) \qquad B,\Gamma''|\Delta'' \Rightarrow^* C}{\Gamma, \Gamma', \Gamma''|\Delta, \Delta', \Delta'' \Rightarrow^* C}\,(\mathrm{Cut}-)$$

Now we can first apply the I.H. to the (Cut+) to get a cut-free derivation of $\Gamma, \Gamma'|\Delta, \Delta' \Rightarrow^- B$; then we can apply the I.H. to the (Cut−) because it has a lower grade. Other cases are similarly argued. □

Lemma 5.8 is enough to establish the cut-eliminability of the systems.

Theorem 5.9 (cut-elimination). Let $\dagger \in \{ab, po, wn\}$. Then $\vdash_{s\dagger} \Gamma|\Delta \Rightarrow^* \Pi$ if and only if $\vdash_{s\dagger}^{cf} \Gamma|\Delta \Rightarrow^* \Pi$.

Proof. From Lemma 5.8, it is possible to transform a derivation with (Cut+) and (Cut−) into a cut-free one by removing, step by step, one of the uppermost instances of (Cut+) or (Cut−). □

5.3 Properties of Cut-free Systems

An immediate corollary of Theorem 5.9 is the following *subformula property* of $\mathbf{SC}^{ab}$:

Corollary 5.10 (subformula property). If $\vdash_{sab} \Gamma|\Delta \Rightarrow^* \Pi$ then there is a derivation of the b-sequent in which all formulas are a subformula of $\Gamma \cup \Delta \cup \Pi$.

Proof. By inspection of the rules in $\mathbf{SC}^{ab}$-(Cut). □

On the other hand, the same argument does not show that the systems $\mathbf{SC}^{ab}_{po}$ and $\mathbf{SC}^{ab}_{wn}$ enjoy the subformula property: they have rules which can eliminate a formula, which leads to its not occurring in the endsequent as a subformula. Despite this, we shall see that when it comes to $\mathbf{SC}^{ab}_{po}$, it is always possible to convert any derivation into a derivation in which all formulas occurring also occur in the endsequent as a subformula.

Following the example of *analytic cut* (see e.g. [12, 27]), we shall call an instance of (sPO)/(L⊥−) *analytic*, if the active formula occurs in the conclusion of the rule as a subformula. Our aim here is to eliminate non-analytic instances of the rules that affect the subformula property.

Lemma 5.11. Let N be either a propositional variable or $\bot$. For any derivation of $\Gamma|\Delta \Rightarrow^*$ Π in $\mathbf{SC}_{po}^{ab}$-(Cut), suppose all instances of (sPO)/(L$\bot$−) in the derivation are analytic. Then $N \notin Sub((\Gamma \setminus \{N\}) \cup \Delta \cup \Pi)$ implies there is a derivation of $\Gamma \setminus \{N\}|\Delta \Rightarrow^* \Pi$ in which all instances of (sPO)/(L$\bot$−) are analytic.

Proof. We show by induction on the depth of derivation. If the derivation is an instance of (Ax−):

$$A| \Rightarrow^- A.$$

Then $N \notin Sub((\{A\} \setminus \{N\}) \cup \{A\})$ implies $\{A\} \setminus \{N\} = \{A\}$. So the derivation is the desired derivation of $\{A\} \setminus \{N\}| \Rightarrow^- A$. For (Ax+) and (L$\bot$+), the statement follows trivially.

Suppose the derivation ends with an instance of (LW−):

$$\frac{\Gamma|\Delta \Rightarrow^* \Pi}{A, \Gamma|\Delta \Rightarrow^* \Pi}$$

Then $N \notin Sub((\{A\} \cup \Gamma) \setminus \{N\} \cup \Delta \cup \Pi)$ implies $N \notin Sub(\Gamma \setminus \{N\} \cup \Delta \cup \Pi)$. Hence by the I.H. there is a derivation of $\Gamma \setminus \{N\}|\Delta \Rightarrow^* \Pi$ in which all instances of (sPO)/(L$\bot$−) are analytic. Now if $A \equiv N$ then this is a desired derivation of $(\{A\} \cup \Gamma) \setminus \{N\}|\Delta \Rightarrow^* \Pi$. If on the other hand $A \not\equiv N$, then apply (LW−) to obtain a desired derivation.

Suppose the derivation ends with an instance of (L∧−):

$$\frac{A, \Gamma|\Delta \Rightarrow^* \Pi \qquad B, \Gamma|\Delta \Rightarrow^* \Pi}{A \wedge B, \Gamma|\Delta \Rightarrow^* \Pi}$$

Then $N \notin Sub(((\{A \wedge B\} \cup \Gamma) \setminus \{N\}) \cup \Delta \cup \Pi)$ implies $N \notin Sub(((\{A\} \cup \Gamma) \setminus \{N\}) \cup \Delta \cup \Pi)$, $N \notin Sub(((\{B\} \cup \Gamma) \setminus \{N\}) \cup \Delta \cup \Pi)$ and in particular $N \not\equiv A, B$. Thus by the I.H. there are derivations of $(\{A\} \cup \Gamma) \setminus \{N\}|\Delta \Rightarrow^* \Pi$ and $(\{B\} \cup \Gamma) \setminus \{N\}|\Delta \Rightarrow^* \Pi$ in which all instances of (sPO)/(L$\bot$−) are analytic. Now, because $N \not\equiv A, B$ we can apply (L∧−) to obtain a desired derivation of $(\{A \wedge B\} \cup \Gamma) \setminus \{N\}|\Delta \Rightarrow^* \Pi$ (note $N \not\equiv A \wedge B$ since it is not a compound formula).

Suppose the derivation ends with an instance of (L→−):

$$\frac{\Gamma|\Delta \Rightarrow^+ A \qquad B, \Gamma'|\Delta' \Rightarrow^* \Pi}{A \to B, \Gamma, \Gamma'|\Delta, \Delta' \Rightarrow^* \Pi}$$

Then $N \notin Sub(((\{A \to B\} \cup \Gamma \cup \Gamma') \setminus \{N\}) \cup \Delta \cup \Delta' \cup \Pi)$ implies $N \notin Sub((\Gamma \setminus \{N\}) \cup \Delta \cup \{A\})$ and $N \notin Sub(((\{B\} \cup \Gamma') \setminus \{N\}) \cup \Delta' \cup \Pi)$. Hence by the I.H. we have the derivations of $\Gamma \setminus \{N\}|\Delta \Rightarrow^+ A$ and $(\{B\} \cup \Gamma') \setminus \{N\}|\Delta' \Rightarrow^* \Pi$. Noting $B, A \to B \not\equiv N$, we can apply (L→−) to obtain $(\{A \to B\} \cup \Gamma \cup \Gamma') \setminus \{N\}|\Delta, \Delta' \Rightarrow^* \Pi$.

Suppose the derivation ends with an instance of (L∼−):

$$\frac{\Gamma|\Delta, A \Rightarrow^* \Pi}{\sim A, \Gamma|\Delta \to^* \Pi}$$

Then $N \notin Sub(((\{{\sim}A\} \cup \Gamma) \setminus \{N\}) \cup \Delta \cup \Pi)$ implies $N \notin Sub((\Gamma \setminus \{N\}) \cup \Delta \cup \{A\} \cup \Pi)$. So there is a derivation of $\Gamma \setminus \{N\}|\Delta, A \Rightarrow^* \Pi$ in which all instances of (sPO)/(L⊥−) are analytic. Apply (L$\sim$−) to obtain a desired derivation of $(\{{\sim}A\} \cup \Gamma) \setminus \{N\}|\Delta \Rightarrow^* \Pi$.

Suppose the derivation ends with an instance of (L$\sim$+):

$$\frac{A, \Gamma|\Delta \Rightarrow^* \Pi}{\Gamma|\Delta, {\sim}A \Rightarrow^* \Pi}$$

Then $N \notin Sub((\Gamma \setminus \{N\}) \cup \Delta \cup \{{\sim}A\} \cup \Pi)$ implies $N \notin Sub(((\{A\} \cup \Gamma) \setminus \{N\}) \cup \Delta \cup \Pi)$. Hence by the I.H. there is a derivation of $(\{A\} \cup \Gamma) \setminus \{N\}|\Delta \Rightarrow^* \Pi$ wherein all instances of (sPO)/(L⊥−) are analytic. Noting $A \not\equiv N$, we can apply (L$\sim$+) to obtain a desired derivation of $(\Gamma) \setminus \{N\}|\Delta, {\sim}A \Rightarrow^* \Pi$.

Suppose the derivation ends with an instance of (sPO):

$$\frac{p, \Gamma|\Delta \Rightarrow^* \qquad \Gamma|\Delta, p \Rightarrow^*}{\Gamma|\Delta \Rightarrow^*}$$

Then by assumption the instance must be analytic, and $N \notin Sub((\Gamma \setminus \{N\}) \cup \Delta)$ implies $N \notin Sub(((\{p\} \cup \Gamma) \setminus \{N\}) \cup \Delta)$. Now if $N \equiv p$, then by the I.H. there is a derivation of $(\{p\} \cup \Gamma) \setminus \{N\}|\Delta \Rightarrow^*$, and this derivation is also a desired derivation of $\Gamma \setminus \{N\}|\Delta \Rightarrow^*$. If $N \not\equiv p$, then $N \notin Sub((\Gamma \setminus \{N\}) \cup \Delta \cup \{p\})$ as well. So we have derivations of $(\{p\} \cup \Gamma) \setminus \{N\}|\Delta \Rightarrow^*$ and $\Gamma \setminus \{N\}|\Delta, p \Rightarrow^*$. As $N \not\equiv p$, we can apply (sPO) to obtain a desired derivation of $\Gamma \setminus \{N\}|\Delta \Rightarrow^*$: note in particular that the application remains analytic because $N \not\equiv p$.

Other cases can be argued analogously. □

Theorem 5.12. If $\vdash^{cf}_{spo} \Gamma|\Delta \Rightarrow^* \Pi$, then there is a derivation of the b-sequent in which all instances of (sPO) and (L⊥−) are analytic.

Proof. Given a derivation of $\Gamma|\Delta \Rightarrow^* \Pi$, we consider a topmost instance of non-analytic (sPO) or (L⊥−). By definition, the active formula N in its (left) premise $N, \Gamma'|\Delta' \Rightarrow^\dagger$ does not occur in the conclusion $\Gamma'|\Delta' \Rightarrow^\dagger$ as a subformula. This means we can apply Lemma 5.11 to obtain a subderivation of $\Gamma'|\Delta' \Rightarrow^\dagger$ in which all instances of (sPO)/(L⊥+) are analytic. This reduces the number of non-analytic (sPO)/(L⊥−) in the new overall derivation, and so we can eliminate all instances by repeating the process. □

Hence we can conclude that the subformula property holds for $\mathbf{SC}^{ab}_{po}$ as well.

Corollary 5.13 (subformula property). If $\vdash_{spo} \Gamma|\Delta \Rightarrow^* \Pi$ then there is a derivation of the b-sequent in which all formulas are a subformula of $\Gamma \cup \Delta \cup \Pi$.

Proof. By inspection on the rules in $\mathbf{SC}^{ab}_{po}$-(Cut) restricted with analytic instances of (sPO) and (L⊥−). □

Next, let us move on to consider the constructivity of the systems, conceived here by means of the disjunction property. We begin with the cases for $\mathbf{SC}^{ab}$ and $\mathbf{SC}^{ab}_{po}$, where, analogously to the case for intuitionistic logic [27], the property holds as a consequence of cut-elimination.

Corollary 5.14 (disjunction property). Let Γ and Δ be finite sets of formulas such that there is no occurrence of $\{\wedge, \sim\}$ in Γ and no occurrence of $\{\vee, \sim\}$ in Δ. Then for $\dagger \in \{ab, po\}$, $\vdash_{s\dagger} \Gamma|\Delta \Rightarrow^+ A \vee B$ implies $\vdash_{s\dagger} \Gamma|\Delta \Rightarrow^+ A$ or $\vdash_{s\dagger} \Gamma|\Delta \Rightarrow^+ B$.

Proof. Suppose that a cut-free derivation of such a b-sequent is given. Then following a path in the derivation upwards, we can construct a finite sequence $s_0, \ldots, s_n$ of b-sequents such that s_0 is $\Gamma|\Delta \Rightarrow^+ A \vee B$, s_{i+1} is the premise of s_i whose succedent is $A \vee B$, and s_n does not have a premise whose succedent is $A \vee B$. Note that the choice of s_{i+1} is uniquely made, as we do not meet an application of $(L\wedge-)$ nor $(L\vee+)$.

Then s_n cannot be an instance of $(Ax+)$ because Δ does not contain a disjunctive formula. It is not difficult to similarly check other rules to see that the rule applied to obtain s_n must be either $(RW+)$ or $(R\vee+)$. Consider the latter case, and assume that the succedent in the premise is A. Then take the premise as s'_n. The we can successively define new b-sequents s'_i whose only difference is that the succedents are A. In particular, each s'_i for $i < n$ is obtained by an application of the same rule. This gives a desired derivation of $\Gamma|\Delta \Rightarrow^+ A$. It is analogously argued when the rule applied is $(RW+)$. $\square$

The constructible falsity property of the systems (with the same class of antecedent formulas) then follows immediately from the disjunction property. On the other hand, the general disjunction property does not hold with respect to $\mathbf{SC}^{ab}_{wn}$, when conceived with the same class of formulas in the antecedent.

Proposition 5.15. $\vdash_{swn} \quad |\neg(p \wedge q) \Rightarrow^+ \sim p \vee \sim q$ but $\nvdash_{swn} \quad |\neg(p \wedge q) \Rightarrow^+ \sim p$ and $\nvdash_{swn} |\neg(p \wedge q) \Rightarrow^+ \sim q$

Proof. The first part is verified by the next derivation:

$$
\dfrac{\dfrac{\dfrac{\dfrac{\dfrac{p| \Rightarrow^- p}{p|\neg(p \wedge q) \Rightarrow^- p}}{p|\neg(p \wedge q) \Rightarrow^+ \sim p}}{p|\neg(p \wedge q) \Rightarrow^+ \sim p \vee \sim q} \quad \overline{\overline{q|\neg(p \wedge q) \Rightarrow^+ \sim p \vee \sim q}}}{p \wedge q|\neg(p \wedge q) \Rightarrow^+ \sim p \vee \sim q} \quad \dfrac{\dfrac{|p \wedge q \Rightarrow^+ p \wedge q \quad |\bot \Rightarrow^+}{|p \wedge q, \neg(p \wedge q) \Rightarrow^+}}{}\text{ (sWN)}}{|\neg(p \wedge q) \Rightarrow^+ \sim p \vee \sim q}
$$

As for the second part, by Proposition 5.3 and Corollary 3.6, it suffices to provide counter-models for $\vDash_{wn} \neg(p \wedge q) \Rightarrow \sim p$ and $\vDash_{wn} \neg(p \wedge q) \Rightarrow \sim q$. For the former, let $\mathcal{M} = ((W, \leq), \mathcal{V})$ be a $\mathbf{C}^{ab}_{wn}$-model such that $W = \{w\}$, $\leq = \{(w, w)\}$, $\mathcal{V}^+(p) = \mathcal{V}^-(q) = W$ and $\mathcal{V}^-(p) = \mathcal{V}^+(q) = \mathcal{V}^-(\bot) = \emptyset$. Otherwise, $\mathcal{V}^+$ and $\mathcal{V}^-$ are defined according to the equivalences in Definition 3.1. Then we can inductively check $w \Vdash^+_{wn} A$ or $w \Vdash^-_{wn} A$ for all A; thus (Weak Negation) is satisfied. Now clearly, $w \Vdash^+_{wn} \neg(p \wedge q)$ but $w \nVdash^+ \sim p$. The latter case is analogous. $\square$

It therefore appears that $\mathbf{C}^{ab}_{wn}$ does not enjoy the same level of constructivity[9] as $\mathbf{C}^{ab}$ and $\mathbf{C}^{ab}_{po}$. This suggest that $\mathbf{C}^{ab}_{wn}$ may not be fully acceptable[10] to an intuitionistic logician similarly to the case for $\mathbf{C3}$.

6 Mixed Constructible Falsity

In this section, we shall look at a further interaction between $\neg$ and $\sim$ which holds in many systems in the vicinity of $\mathbf{C}^{ab}$. We first introduce the notion of *reduced formula*, commonly used for systems with constructible falsity since Nelson [21], as a preliminary notion.

Definition 6.1. For each formula A in $\mathcal{L}$, we define its *reduced formula* $r(A)$ by the following clauses.

$$r(p) = p. \qquad\qquad r(\sim p) = \sim p.$$
$$r(\bot) = \bot. \qquad\qquad r(\sim\bot) = \sim\bot.$$
$$r(A \wedge B) = r(A) \wedge r(B). \qquad r(\sim(A \wedge B)) = r(\sim A) \vee r(\sim B).$$
$$r(A \vee B) = r(A) \vee r(B). \qquad r(\sim(A \vee B)) = r(\sim A) \wedge r(\sim B).$$
$$r(A \to B) = r(A) \to r(B). \qquad r(\sim(A \to B)) = r(A) \to r(\sim B).$$
$$r(\sim\sim A) = r(A).$$

We shall set $r(\Gamma) := \{r(A) : A \in \Gamma\}$. Reduced formulas for $\mathbf{C}$ are already discussed by Wansing [33]. Some standard properties shown therein hold in the current setting as well:

Lemma 6.2. The following statements hold.

(i) $\vdash_{hab} A \leftrightarrow r(A)$.

(ii) $\vdash_{hab} r(\sim A) \leftrightarrow \sim r(A)$.

(iii) $\vdash_{hab} \sim A \leftrightarrow \sim r(A)$.

Proof. We shall show (i) and (ii) by induction on the complexity of A. Here we shall treat the cases where $A \equiv \sim(B \to C)$.

For (i), $r(\sim(B \to C)) \equiv r(B) \to r(\sim C)$. By the I.H. $\vdash_{hab} B \leftrightarrow r(B)$ and $\vdash_{hab} \sim C \leftrightarrow r(\sim C)$. The equivalence then follows from (NI).

For (ii), $r(\sim\sim(B \to C)) \equiv r(B) \to r(C)$ and $\sim r(\sim(B \to C)) \equiv \sim(r(B) \to r(\sim C))$. By the I.H. $\vdash_{hab} r(\sim\sim C) \leftrightarrow \sim r(\sim C)$; so from this and (NI) the statement holds.

Now, $\vdash_{hab} \sim A \leftrightarrow r(\sim A)$ from (i) and $\vdash_{hab} r(\sim A) \leftrightarrow \sim r(A)$ from (ii); so (iii) follows. $\square$

Next we introduce a class of formulas in $\mathcal{L}$.

---

[9]We do not know if $\mathbf{SC}^{ab}_{wn}$ enjoys the disjunction property with the empty antecedent.

[10]It might be argued that $\mathbf{C}$ is already disfavourable for a similar reason: $\Gamma \vdash A \vee B$ (where Γ is disjunction-free) implies $\Gamma \vdash A$ or $\Gamma \vdash B$ in intuitionistic logic but not in $\mathbf{C}$. In this case, however, the two logics have different languages, so it is less clear that we can draw the conclusion that $\mathbf{C}$ is less constructive than intuitionistic logic.

Definition 6.3. Let $\mathcal{F}$ be a class of formulas in $\mathcal{L}$ given by the next clauses.

$$F ::= \bot \mid (A \wedge F) \mid (F \wedge A) \mid (F \vee F) \mid (A \to F).$$

With respect to this class, we have the following couple of lemmas.

Lemma 6.4. Let $\dagger \in \{ab, po\}$. If $\vdash_{s\dagger} \Gamma|\Delta \Rightarrow^*$, then $\bigwedge r(\sim\Gamma) \wedge \bigwedge r(\Delta) \in \mathcal{F}$.

Proof. We show by induction on the depth of derivation. By Theorem 5.9, it suffices to consider the cut-free derivations. Also we may check via soundness that the antecedent is non-empty.

The derivation cannot be an instance of (Ax$-$) or (Ax$+$). If it is an instance of (L$\bot$+), then $r(\bot) \equiv \bot \in \mathcal{F}$.

Otherwise, the derivation ends with an instance of a left rule or (sPO). If it ends with an instance of (LW$-$):

$$\frac{\Gamma|\Delta \Rightarrow^*}{A, \Gamma|\Delta \Rightarrow^*}$$

then by the I.H. $\bigwedge r(\sim\Gamma) \wedge \bigwedge r(\Delta) \in \mathcal{F}$ Hence $r(\sim A) \wedge \bigwedge r(\sim\Gamma) \wedge \bigwedge r(\Delta) \in \mathcal{F}$, as required. The case for (LW$+$) is analogous.

If the derivation ends with an instance of (L$\wedge-$):

$$\frac{A, \Gamma|\Delta \Rightarrow^* \qquad B, \Gamma|\Delta \Rightarrow^*}{A \wedge B, \Gamma|\Delta \Rightarrow^*}$$

then by the I.H. $r(\sim A) \wedge \bigwedge r(\sim\Gamma) \wedge \bigwedge r(\Delta) \in \mathcal{F}$ and $r(\sim B) \wedge \bigwedge r(\sim\Gamma) \wedge \bigwedge r(\Delta) \in \mathcal{F}$. Now if there is $C \in \sim\Gamma \cup \Delta$ such that $r(C) \in \mathcal{F}$, then the statement follows. Otherwise, it must be that $r(\sim A), r(\sim B) \in \mathcal{F}$ Hence $r(\sim(A \wedge B)) \equiv r(\sim A) \vee \sim(B) \in \mathcal{F}$. Hence the statement follows in all cases. The case for (L$\vee+$) is analogous.

If the derivation ends with an instance of (L$\wedge+$):

$$\frac{\Gamma|\Delta, A_i \Rightarrow^*}{\Gamma|\Delta, A_1 \wedge A_2 \Rightarrow^*}$$

then by the I.H. $\bigwedge r(\sim\Gamma) \wedge \bigwedge r(\Delta) \wedge r(A_i) \in \mathcal{F}$. Hence $\bigwedge r(\sim\Gamma) \wedge \bigwedge r(\Delta) \wedge r(A_1 \wedge A_2) \in \mathcal{F}$. The case for (L$\vee-$) is analogous.

If the derivation ends with an instance of (L$\to-$):

$$\frac{\Gamma|\Delta \Rightarrow^+ A \qquad B, \Gamma'|\Delta' \Rightarrow^*}{A \to B, \Gamma, \Gamma'|\Delta, \Delta' \Rightarrow^*}$$

then by the I.H. $r(\sim B) \wedge \bigwedge r(\sim\Gamma') \wedge \bigwedge r(\Delta') \in \mathcal{F}$. If there is $C \in \sim\Gamma' \cup \Delta'$ such that $r(C) \in \mathcal{F}$, then the statement follows. Otherwise, $r(\sim B) \in \mathcal{F}$, so $r(\sim(A \to B)) \equiv r(A) \to r(\sim B) \in \mathcal{F}$. Hence the statement follows in both cases. The case for (L$\to+$) is analogous.

If the derivation ends with an instance of (L$\sim-$):

$$\frac{\Gamma|\Delta, A \Rightarrow^*}{\sim A, \Gamma|\Delta \Rightarrow^*}$$

then by the I.H. $\bigwedge r(\sim\Gamma) \wedge \bigwedge r(\Delta) \wedge r(A) \in \mathcal{F}$. Now if $r(A) \in \mathcal{F}$, then $r(\sim\sim A) \in \mathcal{F}$ and so the statement follows. Otherwise, the statement follows from the I.H.. The case for (L$\sim$+) also follows trivially by the I.H..

If the derivation ends with an instance of (sPO):

$$\frac{p, \Gamma|\Delta \Rightarrow^* \qquad \Gamma|\Delta, p \Rightarrow^*}{\Gamma|\Delta \Rightarrow^*}$$

Then by the I.H. $r(\sim p) \wedge \bigwedge r(\sim\Gamma) \wedge r(\Delta) \in \mathcal{F}$ and $\bigwedge r(\sim\Gamma) \wedge r(\Delta) \wedge r(p) \in \mathcal{F}$. Then because $\sim p, p \notin \mathcal{F}$, there must be $A \in \sim\Gamma \cup \Delta$ such that $r(A) \in \mathcal{F}$. Thus $\bigwedge r(\sim\Gamma) \wedge r(\Delta) \in \mathcal{F}$. The case for (L$\perp$−) is analogous. $\qquad\square$

Lemma 6.5. If $A \in \mathcal{F}$ then $\vdash_{hwn} \sim A$.

Proof. By induction on the construction of formulas in $\mathcal{F}$. If $A \equiv \perp$, then $\vdash_{hwn} \sim\perp$ follows from (WN).

If $A \equiv B \wedge F$, then by the I.H. $\vdash_{hwn} \sim F$. Hence $\vdash_{hwn} \sim(B \wedge F)$ by (DI) and (NC). The case $A \equiv F \wedge B$ is analogous.

If $A \equiv F_1 \vee F_2$, then by the I.H. $\vdash_{hwn} \sim F_1$ and $\vdash_{hwn} \sim F_2$. Hence $\vdash_{hwn} \sim(F_1 \vee F_2)$ by (CI) and (ND).

If $A \equiv B \to F$, then by the I.H. $\vdash_{hwn} \sim F$. Hence $\vdash_{hwn} \sim(B \to F)$ by (K) and (NI). $\quad\square$

The lemmas allow us to establish the next relationship between $\neg$ and $\sim$. (The first item is in fact obvious from (WN); an alternative proof is given here for the interest of a posterior remark.)

Theorem 6.6. The following statements hold.

(i) If $\vdash_{hwn} \neg A$ then $\vdash_{hwn} \sim A$.

(ii) If $\vdash_{hwn} \neg(A \wedge B)$ then $\vdash_{hwn} \sim A$ or $\vdash_{hwn} \sim B$.

Proof. For (i), by Proposition 4.5, if $\vdash_{hwn} \neg A$ then $\vdash_{hpo} \neg A$. Thus by Proposition 5.3, $\vdash_{spo} \ | \ \Rightarrow^+ \neg A$; by (Cut+), $\vdash_{spo} |A \Rightarrow^+$. Hence $r(A) \in \mathcal{F}$ by Lemma 6.4 and so $\vdash_{hwn} \sim r(A)$ by Lemma 6.5. Finally, Lemma 6.2 (iii) implies the desired conclusion.

For (ii), like in (i) if $\vdash_{hwn} \neg(A \wedge B)$ then $r(A \wedge B) \equiv r(A) \wedge r(B) \in \mathcal{F}$. This implies that either $r(A) \in \mathcal{F}$ or $r(B) \in \mathcal{F}$. Then we follow the same path to conclude that $\vdash_{hwn} \sim A$ or $\vdash_{hwn} \sim B$. $\qquad\square$

Therefore in $\mathbf{C}_{wn}^{ab}$, we obtain a sort of 'mixed constructible falsity' property, where the witness for an intuitionistically negated conjunction is given in terms of constructible falsity. This property may be seen to offer an alternative answer for intuitionistic logicians to the failure of the constructible falsity property for intuitionistic negation. Instead of introducing an alternative notion of negation which replaces intuitionistic negation (as happens in **N4**), the connexive constructible falsity of $\mathbf{C}_{wn}^{ab}$ complements intuitionistic negation by becoming a witness of an intuitionistically negated conjunction. In this specific sense, **C**-style systems with the property might be called *more intuitionistic* than **N4**-style systems.

Remark 6.7. It is easily seen from the proof of the above theorem that the same properties can be shown with respect to $\mathbf{CN}^\perp$ and any intermediate system which falls under the scope of Proposition 4.5. Moreover, consider the ([9] style) variants of $\mathbf{C}^{ab}$ and $\mathbf{C}^{ab}_{po}$ in which $\sim |$ is added as an axiom schema, and the corresponding sequent calculi with an additional axiom $| \Rightarrow^- \perp$. It is not difficult to observe that the additional rule does not affect the cut-elimination and Lemma 6.5. Hence the properties of Theorem 6.6 hold with respect to these variants as well.

7 Concluding Remarks

The question that motivated our enquiry is how an intuitionistic logician can make sense of **C**-style connexive constructible falsity, and whether there is a related system in which it is made more understandable by relating it with intuitionistic negation. We in particular looked at two candidates $\mathbf{C}^{ab}_{po}$ and $\mathbf{C}^{ab}_{wn}$.

Having looked at their properties, we may ask which one is to be preferred. Here it seems $\mathbf{C}^{ab}_{po}$ is largely more advantageous, because it has a better behaviour in the semantics (Proposition 3.4), less controversial status on the falsity of intuitionistic negation (Proposition 4.2), a subformula calculus (Corollary 5.13) and better constructivity (Corollary 5.14). Moreover, it shows a good property for investigating provable contradictions constructively (Corollary 4.6), while staying close to **C** (Proposition 4.8). We would therefore suggest that this could be a system that satisfies an intuitionistic logician enough, both in terms of its comprehensibility[11] and its formal behaviours.

In comparison, $\mathbf{C}^{ab}_{wn}$ fares not as well as $\mathbf{C}^{ab}_{po}$ in many of the above-mentioned aspects, and the less satisfactory constructive status may be particularly worrying for an intuitionistic logician. Nonetheless, its satisfaction of 'the mixed constructible falsity' property can offer an independent motivation for the system. Since some of its disadvantages may well be rectified (e.g. by a subformula calculus or the disjunction property with the empty antecedent), further investigations can offer an improved evaluation.

Lastly, however, we would like to point out that there is another system that can potentially meet the expectation of an intuitionistic logician. It is the system $\mathbf{C}^\perp$ in [9], i.e. $\mathbf{C}^{ab}$ with an additional axiom schema $\sim\perp$. As we discussed in Remark 6.7, in this system (WN) holds in *the rule form* (i.e. Theorem 6.6 (i)). This relationship between intuitionistic negation and constructible falsity may be enough for an intuitionistic logician to have an adequate understanding of the latter concept. Therefore it seems, from this perspective, the acceptability of $\sim\neg A$ as a theorem and the reading of $\perp$ as falsehood can have a noticeable influence on the preference of intuitionistic logicians.

---

[11]Admittedly, the double negation can make the schema more difficult to makes sense even though the inner $A \vee \sim A$ is readily understandable. However, it is our (perhaps idealised) supposition that intuitionistic logicians *do* understand all intuitionistic connectives; so the presence of the double negation does not pose an issue for their comprehension.

References

[1] Ahmad Almukdad and David Nelson. Constructible falsity and inexact predicates. *The Journal of Symbolic Logic*, 49(1):231–233, 1984.

[2] Alessandro Avellone, Camillo Fiorentini, Guido Fiorino, and Ugo Moscato. A space efficient implementation of a tableau calculus for a logic with a constructive negation. In *International Workshop on Computer Science Logic*, pages 488–502. Springer, 2004.

[3] Katalin Bimbó. *Proof theory: Sequent calculi and related formalisms*. CRC Press, 2014.

[4] John Cantwell. The logic of conditional negation. *Notre Dame Journal of Formal Logic*, 49(3):245–260, 2008.

[5] Sergey Drobyshevich. Tarskian consequence relations bilaterally: some familiar notions. *Synthese*, 198(Suppl 22):5213–5240, 2021.

[6] Paul Égré, Lorenzo Rossi, and Jan Sprenger. De Finettian logics of indicative conditionals part I: Trivalent semantics and validity. *Journal of Philosophical Logic*, 50:187–213, 2021.

[7] Paul Égré, Lorenzo Rossi, and Jan Sprenger. De Finettian logics of indicative conditionals part II: Proof theory and algebraic semantics. *Journal of Philosophical Logic*, 50:215–247, 2021.

[8] Paul Egré and Hans Rott. The Logic of Conditionals. In Edward N. Zalta, editor, *The Stanford Encyclopedia of Philosophy*. Metaphysics Research Lab, Stanford University, Winter 2021 edition, 2021.

[9] Davide Fazio and Sergei P. Odintsov. An algebraic investigation of the connexive logic C. *Studia Logica*, pages 1–31, 2023.

[10] Thomas Macaulay Ferguson. On arithmetic formulated connexively. *IFCoLog Journal of Logics and their Applications*, 3(3):357–376, 2016.

[11] Ichiro Hasuo and Ryo Kashima. Kripke completeness of first-order constructive logics with strong negation. *Logic Journal of IGPL*, 11(6):615–646, 2003.

[12] Andrzej Indrzejczak. *Sequents and Trees*. Birkhäuser, 2021.

[13] Vadim Anatolyevich Jankov. The calculus of the weak "law of excluded middle". *Mathematics of the USSR-Izvestiya*, 2(5):997–1004, 1968.

[14] Norihiro Kamide and Heinrich Wansing. Proof theory of Nelson's paraconsistent logic: A uniform perspective. *Theoretical Computer Science*, 415:1–38, 2012.

[15] Norihiro Kamide and Heinrich Wansing. *Proof theory of N4-related paraconsistent logics*. College Publications London, 2015.

[16] Ori Lahav and Arnon Avron. A unified semantic framework for fully structural propositional sequent systems. *ACM Transactions on Computational Logic (TOCL)*, 14(4):1–33, 2013.

[17] Andrei Andreyevich Markov. Constructive logic (in russian). *Russian Mathematical Survey*, 5(3-37):187–188, 1950.

[18] Storrs McCall. Connexive implication. *The Journal of Symbolic Logic*, 31(3):415–433, 1966.

[19] Pierangelo Miglioli, Ugo Moscato, Mario Ornaghi, and Gabriele Usberti. A constructivism based on classical truth. *Notre Dame Journal of Formal Logic*, 30(1):67–90, 1989.

[20] Sara Negri and Jan von Plato. *Structural Proof Theory*. Cambridge University Press, 2001.

[21] David Nelson. Constructible falsity. *The Journal of Symbolic Logic*, 14(1):16–26, 1949.

[22] Sergei P. Odintsov. The class of extensions of Nelson's paraconsistent logic. *Studia Logica*, 80(2):291–320, 2005.

[23] Sergei P. Odintsov. *Constructive Negations and Paraconsistency: A Categorical Approach to L-fuzzy Relations*. Springer, 2008.

[24] Grigory K. Olkhovikov. On the completeness of some first-order extensions of C. *Journal of Applied Logics*, 10(1):57–114, 2023.

[25] Hitoshi Omori and Heinrich Wansing. *New Essays on Belnap-Dunn Logic*. Springer, 2019.

[26] Hitoshi Omori and Heinrich Wansing. An extension of connexive logic C. In N. Olivietti, R. Verbrugge, S. Negri, and G. Sandu, editors, *Advances in Modal Logic 13*, pages 503–522. College Publications, 2020.

[27] Hiroakira Ono. *Proof Theory and Algebra in Logic*. Springer, Singapore, 2019.

[28] Dag Prawitz. *Natural deduction: A proof-theoretical study*. Courier Dover Publications, 2006.

[29] Ian Rumfitt. 'Yes and No'. *Mind*, 109(436):781–823, 2000.

[30] Franz von Kutschera. Ein verallgemeinerter Widerlegungsbegriff für Gentzenkalküle. *Archiv für mathematische Logik und Grundlagenforschung*, 12:104–118, 1969.

[31] Nikolai Nikolaevich Vorob'ev. A constructive calculus of statements with strong negation. (in russian). *Trudy Matematicheskogo Instituta imeni V.A. Steklova*, 72:195–227, 1964.

[32] Heinrich Wansing. *The logic of information structures*, volume 681 of *Lecture Notes in Computer Science*. Springer, 1993.

[33] Heinrich Wansing. Connexive modal logic. In Renate Schmidt, Ian Pratt-Hartmann, Mark Reynolds, and Heinrich Wansing, editors, *Advances in Modal Logic*, volume 5, pages 387–399. College Publications, 2004.

[34] Heinrich Wansing. Natural deduction for bi-connexive logic and a two-sorted typed λ-calculus. *IfCoLog Journal of Logics and their Applications*, 3(3):413–439, 2016.

[35] Heinrich Wansing. Connexive Logic. In Edward N. Zalta, editor, *The Stanford Encyclopedia of Philosophy*. Metaphysics Research Lab, Stanford University, Summer 2023 edition, 2023.

[36] Heinrich Wansing and Sara Ayhan. Logical multilateralism. *Journal of Philosophical Logic*, pages 1–34, 2023.

[37] Heinrich Wansing, Hitoshi Omori, and Thomas Macaulay Ferguson. The tenacity of connexive logic: Preface to the special issue. *IfCoLog Journal of Logics and their Applications*, 3(3):279–296, 2016.

Received June 2023

ESEMIHOOPS

MIN LIU

School of Mathematics and Statistics, Shandong Normal University, 250014, Jinan, P. R. China

HONGXING LIU[*]

School of Mathematics and Statistics, Shandong Normal University, 250014, Jinan, P. R. China
lhxshanda@163.com

Abstract

In this paper, we introduce the concept of Esemihoops, which is an extension of semihoops and Ehoops. We also give some related properties of these algebras. In addition, we define the notions of ideals, congruences and filters on Esemihoops and study the relations between them. There is a one-to-one correspondence between the set of all ideals and the set of all congruences on a regular Esemihoop. At the same time, it is proved that every proper Esemihoop has at least one maximal filter. For a regular Esemihoop A, A has at least one maximal ideal and every maximal ideal is prime. Furthermore, we prove that a Bosbach state on Esemihoops is a Riečan state. In particular, Bosbach states and Riečan states are consistent on Glivenko Esemihoops. Moreover, we show that there is a one-to-one correspondence between the set of τ-compatible Riečan states on A and the set of Riečan states on $\tau(A)$. Finally, the topology of the set of all prime state filters is established.

Keyword: Esemihoop, semihoop, ideal, congruence, filter, state, internal state

The authors would like to appreciate the anonymous referees' valuable comments which highly improved the paper. This study was funded by Natural Science Foundation of Shandong Province [ZR2022MA043].

[*]Corresponding Author.

1 Introduction

Semihoops are generalizations of hoops, which were initially presented by Bosbach in [5] as a complementary semigroup. The notion of semihoops was defined in [1]. Semihoop is the most basic residuated structure. MV-algebra, BL-algebra and Hoop are all particular cases of it.

In recent years, semihoop theory has been widely studied. Niu, Xin and Zhao [8] defined the notion of ideals on bounded semihoops. In the meantime, they introduced several types of ideals, such as primary ideals, prime ideals and maximal ideals. In addition, they discussed relations between them. Subsequently, He, Wang and Yang [4] investigated semi-MV ideals on bounded semihoops and some properties of it. In [2, 10, 11, 13], the authors introduced various types of filters on semihoops, for example, perfect filters, primary filters, SM-filters, state filters, dual state filters, derivations filters and differential filters. Some significant results were also obtained. Moreover, the study of state theory also plays an important role in semihoops. Many scholars have studied states on semihoops, with the main results in [3, 10, 12].

Dvurečenskij and Zahiri introduced the notion of EMV-algebra in [6], which is an extension of MV-algebras. Then, they defined congruences, ideals and filters on EMV-algebras, and studied the relations between them. It was proved that every proper EMV-algebra as a maximal ideal can be embedded into an EMV-algebra with a top element. Liu [7] defined EBL-algebras, which are generalizations of BL-algebras and EMV-algebras. He proved that there is a one-to-one correspondence between the set of all ideals on an EBL-algebra and the set of all congruences on an EBL-algebra. Xie and Liu [9] introduced Ehoops, which are extensions of hoops. They proved that if an Ehoop A has the double negation property, there is a one-to-one correspondence between the set of all ideals and the set of all congruences on A. In addition, they gave the prime ideal theorem.

Inspired by these papers, we will give the notion of Esemihoops, which is an extension of semihoops and Ehoops. A semihoop with $x \odot (x \to y) = y \odot (y \to x)$ is a hoop. Compared with the Ehoops, Esemihoops does not satisfy $x \wedge a = x \odot a$. Therefore, the proofs of some results on Esemihoops are different from those on Ehoops. The structure of this paper is as follows. In Sect.2, we review some necessary definitions and results of semihoops. In Sect.3, we define Esemihoops and study some basic properties of Esemihoops. In Sect.4, we introduce ideals, congruences and filters of Esemihoops. The relationships between them are discussed. In Sect.5, we give the notions of maximal filters, maximal ideals and prime ideals on Esemihoops. We further investigate the relationship between filters and ideals on Esemihoops. In Sect.6, we present the concept of states on Esemihoops. The relationships between Bosbach states and Riečan states of an Esemihoop are discussed. In Sect.7, we define internal states on Esemihoops and give a one-to-one correspondence between the set of τ-compatible Riečan states on A and the set of Riečan states on $\tau(A)$.

2   Preliminaries

In this section, we shall recall some notions and properties of semihoops, which will be needed in the paper.

Definition 2.1. [1] A semihoop is an algebra $(A, \odot, \rightarrow, \wedge, 1)$ of type $(2, 2, 2, 0)$ satisfying the following conditions:
(SH1) $(A, \wedge, 1)$ is a meet-semilattice with the upper bound 1;
(SH2) $(A, \odot, 1)$ is a commutative monoid;
(SH3) $(x \odot y) \rightarrow z = x \rightarrow (y \rightarrow z)$, for any $x, y, z \in A$.

Let $(A, \odot, \rightarrow, \wedge, 1)$ be a semihoop. We define a relation $\leq$ by $x \leq y \Longleftrightarrow x \rightarrow y = 1$, for all $x, y \in A$. The relation is a partial order on A. A is bounded if there exists $0 \in A$ such that $x \geq 0$ for all $x \in A$. Let A be a bounded semihoop. We define $x^- = x \rightarrow 0$ for all $x \in A$. If $x^{--} = x$ for all $x \in A$, A is called a regular semihoop [4]. If a semihoop A satisfies the condition $x \odot (x \rightarrow y) = y \odot (y \rightarrow x)$ for all $x, y \in A$, A is a hoop. If $S \subseteq A$ and $S \neq \emptyset$, S is said to be a subalgebra of A if it is closed under $\odot$, $\rightarrow$ and $\wedge$.

For two semihoops A and B, a map $f : A \rightarrow B$ is called a semihoop homomorphism if f preserves $\odot$, $\rightarrow$, $\wedge$ and 1. This means that f preserves the top element. That is, $f(1_A) = 1_B$. If A and B are bounded, then $f(0) = 0$.

An equivalence relation θ on a semihoop A is called a congruence if it is compatible with $\odot$, $\rightarrow$ and $\wedge$.

An element a in a monoid $(A, \odot, 1)$ is called an idempotent element if $a \odot a = a$. The set of all idempotent elements of A is denoted by $\mathbf{I}(A)$.

Proposition 2.2. [1, 3] Let $(A, \odot, \rightarrow, \wedge, 1)$ be a semihoop. For all $x, y, z \in A$:
(1) $x \odot y \leq z \Longleftrightarrow x \leq y \rightarrow z$;
(2) $x \odot y \leq x \wedge y$, $x \leq y \rightarrow x$;
(3) $1 \rightarrow x = x$, $x \rightarrow 1 = 1$;
(4) $x \odot (x \rightarrow y) \leq y$, $x \leq (x \rightarrow y) \rightarrow y$;
(5) $x \leq y \Longrightarrow y \rightarrow z \leq x \rightarrow z$, $z \rightarrow x \leq z \rightarrow y$ and $x \odot z \leq y \odot z$;
(6) $((x \rightarrow y) \rightarrow y) \rightarrow y = x \rightarrow y$;
(7) $x \rightarrow (y \rightarrow z) = y \rightarrow (x \rightarrow z)$;
(8) $x \rightarrow y \leq (z \rightarrow x) \rightarrow (z \rightarrow y)$, $x \rightarrow y \leq (y \rightarrow z) \rightarrow (x \rightarrow z)$;
(9) $x \rightarrow y \leq (x \wedge z) \rightarrow (y \wedge z)$, $x \rightarrow y \leq (x \odot z) \rightarrow (y \odot z)$;

(10) $x \to (x \wedge y) = x \to y$;
(11) $x \odot y = x \odot (x \to (x \odot y))$.

Proposition 2.3. [2] Let $(A, \odot, \to, \wedge, 1)$ be a bounded semihoop. Then for all $x, y \in A$, the following conditions hold:
(1) $0^- = 1$, $1^- = 0$;
(2) $x \le x^{--}$, $x^- = x^{---}$;
(3) $x \odot x^- = 0$;
(4) $x \le y \Longrightarrow y^- \le x^-$;
(5) $x \to y \le y^- \to x^-$;
(6) $(x \to y^-)^{--} = x \to y^-$;
(7) $x^{--} \odot y^{--} \le (x \odot y)^{--}$.

Let $(A, \odot, \to, \wedge, 0, 1)$ be a bounded semihoop. We define $\ominus$ by $x \ominus y = x^- \to y$ for all $x, y \in A$.

Definition 2.4. [8] Let $(A, \odot, \to, \wedge, 0, 1)$ be a bounded semihoop and I be a non-empty subset of A. Then I is called an ideal if it satisfies:
(1) if $x, y \in I$, then $x \ominus y \in I$;
(2) $x \le y$ and $y \in I$ imply $x \in I$, for all $x, y \in A$.

An ideal I of A is proper if $1 \notin I$. A proper ideal I of A is maximal, if it is not properly contained in any other proper ideal of A.

Definition 2.5. [2] Let $(A, \odot, \to, \wedge, 1)$ be a semihoop and F be a nonempty subset of A. Then F is called a filter if it satisfies:
(1) if $x, y \in F$, then $x \odot y \in F$;
(2) $x \in F$ and $x \le y$ imply $y \in F$, for all $x, y \in A$.

The equivalent notion of a filter can be defined in a simpler way. Let $(A, \odot, \to, \wedge, 1)$ be a semihoop. A nonempty subset F of A is called a filter if (1) $1 \in F$; (2) $x, x \to y \in F$ imply $y \in F$. A filter F is said to be proper if $F \ne A$. A maximal filter is a proper filter and no proper filter strictly contains it.

Definition 2.6. [10] A semihoop $(A, \odot, \to, \wedge, 1)$ is said to be simple if it has only two filters $\{1\}$ and A.

3 Esemihoops

In this section, we will define Esemihoop, which extends the notions of semihoops and Ehoops. What's more, some basic properties of these algebras are given.

Definition 3.1. An algebra $(A, \wedge, \odot)$ of type $(2,2)$ is called an extended semihoop (abbreviated as Esemihoop) if it satisfies the following conditions:

(ESH1) $(A, \wedge)$ is a $\wedge$-semilattice;

(ESH2) $(A, \odot)$ is a commutative semigroup;

(ESH3) for all $a \in \mathbf{I}(A)$, define $A_a = \{x \in A | x \leq a\}$ for any $x, y \in A_a$, the element $x \rightarrow_a y = max\{z \in A_a | x \odot z \leq y\}$ exists, and $(A_a, \odot, \rightarrow_a, \wedge, a)$ is a semihoop;

(ESH4) $(A, \wedge, \odot)$ has enough idempotent elements, that is, for all $x, y \in A$, there exists $a \in \mathbf{I}(A)$ such that $x, y \leq a$.

Remark 3.2. (1) Form (ESH4), we obtain that there exists $a \in \mathbf{I}(A)$ such that $x_1, x_2, \cdots, x_n \leq a$ for any $x_1, x_2, \cdots, x_n \in A$, where $n \geq 2$.

(2) Let $(A, \odot, \rightarrow, \wedge, 1)$ be a semihoop and $a \in \mathbf{I}(A)$. Suppose that $y \in A$ and $x \in A_a$, then $x \rightarrow (y \wedge a) = (x \rightarrow y) \wedge (x \rightarrow a)$. In fact, we have $x \rightarrow (y \wedge a) = (x \wedge a) \rightarrow (y \wedge a) \geq x \rightarrow y$ by Proposition 2.2. On the other hand, it follows from $y \wedge a \leq y$ that $x \rightarrow (y \wedge a) \leq x \rightarrow y$. These together with $x \rightarrow a = 1$ imply that $x \rightarrow (y \wedge a) = x \rightarrow y = (x \rightarrow y) \wedge (x \rightarrow a)$.

(3) Let $(A, \wedge, \odot)$ be an Esemihoop. For all $a \in \mathbf{I}(A)$, A_a is a semihoop. Thus, for any $x \in A_a$, $x \wedge a = x = x \odot a$ by Definition 2.1.

Remark 3.3. Let $(A, \wedge, \odot)$ be an Esemihoop and $a \in \mathbf{I}(A)$.

(1) For all $x, y \in A$, the relation $\leq$ given by $x \leq y \iff x \wedge y = x$ is a partial order on A.

(2) In the semihoop $(A_a, \odot, \rightarrow_a, \wedge, a)$, there is a partial order defined by $x \leq_a y \iff x \rightarrow_a y = a$ for all $x, y \in A_a$. For all $x, y \in A$, take $a, b \in \mathbf{I}(A)$ such that $x, y \leq a$ and $x, y \leq b$. We assert that $x \leq_a y \iff x \leq y \iff x \leq_b y$. In fact, assuming $x \leq_a y$, we have $x \odot a \leq y$ by (ESH3). Thus, $x \leq y$. Conversely, if $x \leq y$, $x \odot a \leq x \leq y$. Then $a = max\{z \in A_a | x \odot z \leq y\} = x \rightarrow_a y$, which implies $x \leq_a y$. So $x \leq_a y \iff x \leq y$. Similarly, we have $x \leq y \iff x \leq_b y$. This shows that for each $a \in \mathbf{I}(A)$, $\leq$ and $\leq_a$ are coincident on A_a.

Example 3.4. Semihoops are termwise equivalent to Esemihoops with a top element.

Proof. $\Longrightarrow$: Suppose that $(A, \odot, \rightarrow, \wedge, 1)$ is a semihoop. Clearly, (ESH1) and (ESH2) hold by Definition 2.1. Let $a \in \mathbf{I}(A)$. For all $x, y \in A_a$, $x \odot y \leq a \odot a = a$, which implies $x \odot y \in A_a$. From $x \wedge y \leq x \leq a$, we have $x \wedge y \in A_a$. For all $x, y \in A_a$, define $x \rightarrow_a y = (x \rightarrow y) \wedge a$. We assert that $(x \rightarrow y) \wedge a = max\{z \in A_a | x \odot z \leq y\}$. In fact, we have $x \odot ((x \rightarrow y) \wedge a) \leq x \odot (x \rightarrow y) \leq y$. Let $z \in A_a$ such that $x \odot z \leq y$. We have $z \leq x \rightarrow y$ and so $z \leq (x \rightarrow y) \wedge a$. Therefore, $(x \rightarrow y) \wedge a = max\{z \in A_a | x \odot z \leq y\}$. So we can obtain that $x \rightarrow_a y \in A_a$ for all $x, y \in A_a$.

We claim that $(A_a, \odot, \to_a, \wedge, a)$ is a semihoop for each $a \in \mathbf{I}(A)$. Let $x, y, z \in A_a$. It is clear that the conditions (SH1) and (SH2) hold. Moreover, we have

$$
\begin{aligned}
x \to_a (y \to_a z) &= (x \to ((y \to z) \wedge a)) \wedge a \\
&= ((x \to (y \to z)) \wedge (x \to a)) \wedge a \quad \text{(Remark 3.2)} \\
&= (x \to (y \to z)) \wedge ((x \to a) \wedge a) \\
&= ((x \odot y) \to z) \wedge a \\
&= (x \odot y) \to_a z.
\end{aligned}
$$

It follows that (SH3) holds. Thus, $(A_a, \odot, \to_a, \wedge, a)$ is a semihoop. That is, condition (ESH3) holds.

For all $x, y \in A$, there exists $1 \in \mathbf{I}(A)$ such that $x, y \leq 1$. This proves (ESH4). Hence, $(A, \wedge, \odot)$ is an Esemihoop.

$\Longleftarrow$: Let $(A, \wedge, \odot)$ be an Esemihoop. It is easy to know that $(A, \odot, \to_1, \wedge, 1)$ is a semihoop by (ESH3). $\qquad \square$

An Esemihoop $(A, \wedge, \odot)$ without a top element is proper.

Example 3.5. Let $\{(A_i, \odot, \to, \wedge, 0, 1)\}_{i \in I}$ be a family of bounded semihoops and $A = \{f \in \prod_{i \in I} A_i | supp(f) \text{ is finite}\}$, where $supp(f) = \{i \in I | f(i) \neq 0\}$. We define the operations $\odot$ and $\wedge$ on A: for all $f = (f_i)_{i \in I}, g = (g_i)_{i \in I} \in A$,

$$
f \odot g = (f_i \odot g_i)_{i \in I}, \ f \wedge g = (f_i \wedge g_i)_{i \in I}.
$$

It is easy to know that A is closed under $\odot$ and $\wedge$. Then (ESH1) and (ESH2) hold. For all $f, g \in A$,

$$
m = (m_i)_{i \in I} = \begin{cases} 1, & i \in supp(f) \cup supp(g), \\ 0, & otherwise. \end{cases}
$$

is an idempotent element and $f, g \leq m$. Thus, we obtain that the condition (ESH4) holds and $\mathbf{I}(A) = \{(m_i)_{i \in I} | m_i \in \mathbf{I}(A_i)\}$. We define $f \to_m g = (f_i \to g_i)_{i \in I} \wedge m$ for all $m \in \mathbf{I}(A)$. Similar to the proof of Example 3.4, we have $(f_i \to g_i)_{i \in I} \wedge m = max\{h \in A_m | f \odot h \leq g\}$. Next, we only need to prove that $(A_m, \odot, \to_m, \wedge, m)$ is a semihoop for all $m \in \mathbf{I}(A)$. Clearly, we can attain that the condition (SH1) holds. Suppose that $f, g, h \in A_m$. We have $f \odot m = f \wedge m = f$, which implies that the condition (SH2) holds. Moreover, we have

$$
\begin{aligned}
f \to_m (g \to_m h) &= (f_i \to ((g_i \to h_i) \wedge m_i))_{i \in I} \wedge m \\
&= (f_i \to (g_i \to h_i))_{i \in I} \wedge m \\
&= ((f_i \odot g_i) \to h_i)_{i \in I} \wedge m \\
&= (f \odot g) \to_m h.
\end{aligned}
$$

Therefore, $(A, \wedge, \odot)$ is an Esemihoop.

The following example we constructed comes from [7].

Example 3.6. Let A be an Esemihoop and X a nonempty finite set. We denote by A^X the set of all functions from X to A. Define the operations $\odot$ and $\wedge$: for all $f, g \in A^X$ and $x \in X$,

$$(f \odot g)(x) = f(x) \odot g(x), \ (f \wedge g)(x) = f(x) \wedge g(x).$$

Obviously, (ESH1) and (ESH2) hold. For all $x \in X$, there is $a \in \mathbf{I}(A)$ such that $f(x), g(x) \le a$. Let $m_a : X \to a$. We have that m_a is an idempotent element of A^X and $f, g \le m_a$. Thus, A^X has enough idempotent elements. For all $f, g \in A^X_{m_a} = \{f \in A^X | f \le m_a\}$, we define $(f \to_{m_a} g)(x) = f(x) \to_{m_a(x)} g(x)$. It is easy to check that $f(x) \to_{m_a(x)} g(x) = \{h \in A^X_{m_a} | f \odot h \le g\}$ and $A^X_{m_a}$ is a semihoop. Therefore, A^X is an Esemihoop.

Example 3.7. Let A and B be two Esemihoops. Let $A \times B = \{(x_1, x_2) | x_1 \in A, x_2 \in B\}$. The operations $\odot$ and $\wedge$ are defined by: for all $x = (x_1, x_2), y = (y_1, y_2) \in A \times B$,

$$(x_1, x_2) \odot (y_1, y_2) = (x_1 \odot y_1, x_2 \odot y_2), \ (x_1, x_2) \wedge (y_1, y_2) = (x_1 \wedge y_1, x_2 \wedge y_2).$$

There exist $a_1 \in \mathbf{I}(A), a_2 \in \mathbf{I}(B)$ such that $x_1, y_1 \le a_1$ and $x_2, y_2 \le a_2$. Thus, $x, y \le a = (a_1, a_2)$ and a is an idempotent element of $A \times B$. Let $x \to_a y = (x_1 \to_{a_1} y_1, x_2 \to_{a_2} y_2)$. Then, $A \times B$ is an Esemihoop.

Example 3.8. Let $(M, \vee, \wedge, 0, 1, \neg, \to)$ be an R_0 algebra [14] and $(A, \wedge, \odot)$ be an Esemihoop. For all $(x, y) \in M \times A$, there exists $a \in \mathbf{I}(A)$ such that $y \le a$. Thus, $(x, y) \le (1, a)$. Then $M \times A$ is an Esemihoop with the operations defined pointwise.

In the following, we give some basic properties of Esemihoops. These properties will often be used in the paper.

Proposition 3.9. Let $(A, \wedge, \odot)$ be an Esemihoop and $a, b \in \mathbf{I}(A)$ with $a \le b$. For all $x, y, z \in A_a$, we have the following properties:
(1) $x \to_a y = (x \to_b y) \wedge a$;
(2) $x \to_a y \le x \to_b y$;
(3) $a \to_b a = b, a \to_a a = a$;
(4) $(x \to_a y) \to_a z \le (x \to_b y) \to_b z$.

Proof. Similar to the proof of Proposition 3.8 in [9], the proofs of (1), (2) and (3) are direct. In order to prove (4) by (1), we have

$$
\begin{aligned}
(x \to_a y) \to_a z &= (((x \to_b y) \wedge a) \to_b z) \wedge a \\
&\leq (((x \to_b y) \odot a) \to_b z) \wedge a \\
&= (a \to_b ((x \to_b y) \to_b z)) \wedge a \\
&= (a \to_b (a \wedge ((x \to_b y) \to_b z))) \wedge a \quad \text{(Proposition 2.2(10))} \\
&= a \to_a (a \wedge ((x \to_b y) \to_b z)) \\
&= a \wedge ((x \to_b y) \to_b z) \\
&\leq (x \to_b y) \to_b z.
\end{aligned}
$$

$\square$

Let $(A, \wedge, \odot)$ be an Esemihoop with the least element 0. We define the negation $-a$ on A_a as follows: $x^{-a} = x \to_a 0$, for all $a \in \mathbf{I}(A)$ and $x \in A_a$.

Proposition 3.10. Let $(A, \wedge, \odot)$ be an Esemihoop with the least element 0. Then for all $a, b \in \mathbf{I}(A)$ and $x, y, z \leq a, b$, we have $x \odot (y \to_a z) = x \odot (y \to_b z)$. In particular, $x \odot y^{-a} = x \odot y^{-b}$.

Proof. Suppose that $a, b \in \mathbf{I}(A)$ and $x, y, z \leq a$. Since A is an Esemihoop, there exists $c \in \mathbf{I}(A)$ such that $a, b \leq c$. In the semihoop A_c, we get $x \odot (y \to_a z) = x \odot ((y \to_c z) \wedge a) \leq x \odot (y \to_c z)$. At the same time, from $(y \to_c z) \odot a \leq (y \to_c z) \wedge a$, we have

$$
\begin{aligned}
x \odot (y \to_c z) &= (x \odot a) \odot (y \to_c z) \\
&= x \odot ((y \to_c z) \odot a) \\
&\leq x \odot ((y \to_c z) \wedge a) \\
&= x \odot (y \to_a z).
\end{aligned}
$$

Then $x \odot (y \to_a z) = x \odot (y \to_c z)$. Analogously, $x \odot (y \to_b z) = x \odot (y \to_c z)$. Therefore, $x \odot (y \to_a z) = x \odot (y \to_b z)$. In particular, taking $z = 0$, we can obtain that $x \odot y^{-a} = x \odot y^{-b}$. $\square$

Definition 3.11. An Esemihoop A is said to be regular if A has the least element 0 and the semihoop A_a is regular for each $a \in \mathbf{I}(A)$.

Proposition 3.12. Let A be a regular Esemihoop and $a \in \mathbf{I}(A)$ with $x, y \leq a$. We have
(1) $(x \to_a y)^{-a} = x \odot y^{-a}$;
(2) $x^{-a} \to_a y = y^{-a} \to_a x$.

Proof. (1) $(x \to_a y)^{-a} = (x \to_a y^{-a-a})^{-a} = (x \odot y^{-a})^{-a-a} = x \odot y^{-a}$.

(2) $x^{-a} \to_a y = x^{-a} \to_a y^{-a-a} = (x^{-a} \odot y^{-a})^{-a} = y^{-a} \to_a x^{-a-a} = y^{-a} \to_a x$. $\qquad\square$

The following proposition gives an equivalent characterization of Esemihoops.

Proposition 3.13. Let $(A, \wedge, \odot)$ be an algebra of type $(2, 2)$. Then A is an Esemihoop iff
(ESH1) $(A, \wedge)$ is a $\wedge$-semilattice;
(ESH2) $(A, \odot)$ is a commutative semigroup;
(ESH3$'$) for all $x, y \in A$, there is an idempotent a in A with $x, y \in A_a$ and $(A_a, \odot, \to_a, \wedge, a)$ is a semihoop.

Proof. Suppose that $(A, \wedge, \odot)$ is an Esemihoop. Obviously, (ESH1)–(ESH3$'$) hold. Conversely, it is easy to prove that (ESH4) holds. From (ESH3$'$), for each $a \in \mathbf{I}(A)$, there exists $b \in \mathbf{I}(A)$ such that $a \leq b$ and $(A_b, \odot, \to_b, \wedge, b)$ is a semihoop. Just like the proof of Example 3.4, we can get that $x \to_a y = max\{z \in A_a | x \odot z \leq y\} = (x \to_b y) \wedge a$. Hence, this proves (ESH3). $\qquad\square$

Definition 3.14. Let $(A, \wedge, \odot)$ be an Esemihoop. A subset $S \subseteq A$ is a subalgebra of A if it satisfies:
(1) S is closed under $\wedge$ and $\odot$;
(2) For all $a \in \mathbf{I}(A) \cap S$, $S_a = \{x \in S | x \leq a\}$ is a subalgebra of the semihoop $(A_a, \odot, \to_a, \wedge, a)$;
(3) For all $x, y \in S$, there is an idempotent $a \in S$ with $x, y \leq a$.

An Esemihoop homomorphism from an Esemihoop A to an Esemihoop B is a map $f : A \to B$ such that:
(1) f preserves $\wedge$ and $\odot$;
(2) for any $a \in \mathbf{I}(A)$ and $x, y \in A_a$, $f(x \to_a y) = f(x) \to_{f(a)} f(y)$.

For an Esemihoop homomorphism $f : A \to B$, we have that $f(x) \leq f(y)$ if $x \leq y$ for all $x, y \in A$. Indeed, there exists $a \in \mathbf{I}(A)$ such that $x, y \leq a$ and $(A_a, \odot, \to_a, \wedge, a)$ is a semihoop. Thus, we have that $x \to_a y = a$ and so $f(x) \to_{f(a)} f(y) = f(x \to_a y) = f(a)$. We get that $f(x) \leq_{f(a)} f(y)$. That is, $f(x) \leq f(y)$ by Remark 3.3.

Remark 3.15. Every semihoop homomorphism is an Esemihoop homomorphism, but the converse is not necessarily true. In fact, for a semihoop $(A, \odot, \to, \wedge, 1)$ and $1 \neq a \in \mathbf{I}(A)$, $(A_a, \odot, \to_a, \wedge, a)$ is also a semihoop. The inclusion map $f : A_a \to A$ is an Esemihoop homomorphism. As $f(a) = a$ and $f(a) \neq 1$, we get that f is not a semihoop homomorphism.

4 Congruences, ideals and filters

In this section, we will define the notions of congruences, ideals and filters in Esemi-hoops and study the relations between them. The equivalent characterizations of ideals and filters are given. Moreover, we construct congruences on an Esemihoop by ideals and filters. It is proved that there is a one-to-one correspondence between the set of all ideals and the set of all congruences on a regular Esemihoop.

Definition 4.1. Let $(A, \wedge, \odot)$ be an Esemihoop. An equivalence relation θ on A is said to be a congruence if it satisfies:
(1) θ is compatible with $\wedge$ and $\odot$;
(2) for each $a \in \mathbf{I}(A)$, $\theta \cap (A_a \times A_a)$ is a congruence on the semihoop A_a.

Let $(A, \wedge, \odot)$ be an Esemihoop. Suppose that θ is a congruence on A and $A/\theta = \{x/\theta | x \in A\}$. Now we define $\wedge$ and $\odot$ on A/θ as follows:

$$x/\theta \wedge y/\theta = (x \wedge y)/\theta, \; x/\theta \odot y/\theta = (x \odot y)/\theta, \text{ for all } x/\theta, \, y/\theta \in A/\theta.$$

We claim that $(A/\theta, \wedge, \odot)$ is an Esemihoop. By Proposition 3.13, we only need to prove that $((A/\theta)_{a/\theta}, \odot, \rightarrow_{a/\theta}, \wedge, a/\theta)$ is a semihoop.

Define $\theta_a = \theta \cap (A_a \times A_a)$. We have that θ_a is a congruence on the semihoop A_a. Also, $(A_a/\theta_a, \odot, \rightarrow_{a/\theta_a}, \wedge, a/\theta_a)$ is a semihoop with the operations defined by:

$$\alpha/\theta_a \odot \beta/\theta_a = (\alpha \odot \beta)/\theta_a, \; \alpha/\theta_a \rightarrow_{a/\theta_a} \beta/\theta_a = (\alpha \rightarrow_a \beta)/\theta_a.$$

For $x/\theta \in (A/\theta)_{a/\theta}$, we have that $(x \wedge a)/\theta = x/\theta \wedge a/\theta = x/\theta$ and $x \wedge a \in A_a$. Thus, we can assume $x \in A_a$. Clearly, we have $x/\theta \odot (x \rightarrow_a y)/\theta = (x \odot (x \rightarrow_a y))/\theta \leq y/\theta$. If $u/\theta \in (A/\theta)_{a/\theta}$ and $x/\theta \odot u/\theta \leq y/\theta$, we have $(x/\theta \odot u/\theta) \wedge y/\theta = x/\theta \odot u/\theta$. Then $((x \odot u) \wedge y, x \odot u) \in \theta$. Assume $x, y, z \in A_a$. We get $((x \odot u) \wedge y, x \odot u) \in \theta_a$. It follows that $(x \odot u)/\theta_a = ((x \odot u) \wedge y)/\theta_a \leq y/\theta_a$. In the semihoop A_a/θ_a, we get $u/\theta_a \leq x/\theta_a \rightarrow_{a/\theta_a} y/\theta_a = (x \rightarrow_a y)/\theta_a$. Then we have $((x \rightarrow_a y) \wedge u, u) \in \theta_a \subseteq \theta$, which implies that $u/\theta \leq (x \rightarrow_a y)/\theta$. Therefore, we obtain that $x/\theta \rightarrow_{a/\theta} y/\theta = max\{z/\theta \in (A/\theta)_{a/\theta} | x/\theta \odot z/\theta \leq y/\theta\} = (x \rightarrow_a y)/\theta$ for all $x/\theta, y/\theta \in (A/\theta)_{a/\theta}$. Thus, $((A/\theta)_{a/\theta}, \odot, \rightarrow_{a/\theta}, \wedge, a/\theta)$ is a semihoop.

Let $(A, \wedge, \odot)$ be an Esemihoop. We define the map $f : A \rightarrow A/\theta$ by $f(x) = x/\theta$ for all $x \in A$. Then f is an Esemihoop homomorphism. If A has the least element 0, for each $a \in \mathbf{I}(A)$, we define $\ominus_a$ on A as follows: $x \ominus_a y = x^{-a} \rightarrow_a y$ for all $x, y \in A_a$.

Definition 4.2. Let A be an Esemihoop with the least element 0. A nonempty subset I of A is said to an ideal if it satisfies:
(ESI1) $x \ominus_a y \in I$, for all $x, y \in I$;
(ESI2) $x \leq y$ and $y \in I$ imply that $x \in I$.

I is said to be proper if $I \neq A$. The following proposition gives an equivalent characterization of ideals of an Esemihoop.

Proposition 4.3. Let $(A, \wedge, \odot)$ be an Esemihoop with the least element 0 and I be a nonempty subset of A. The following statements are equivalent: for all $x, y \in A$ and $x, y \leq a$, where $a \in \mathbf{I}(A)$,
(1) I is an ideal of A;
(2) $0 \in I$, $x^{-a} \odot y \in I$ and $x \in I \implies y \in I$;
(3) $0 \in I$, $(x^{-a} \rightarrow_a y^{-a})^{-a} \in I$ and $x \in I \implies y \in I$.

Proof. (1) $\implies$ (2) Suppose that I is an ideal of A. From Definition 4.2, we have $0 \in I$. Now, for all $x, y \in A$ and $a \in \mathbf{I}(A)$ with $x, y \leq a$, let $x^{-a} \odot y \in I$ and $x \in I$. Then $x \ominus_a (x^{-a} \odot y) \in I$ and $y \leq x^{-a} \rightarrow_a (x^{-a} \odot y) = x \ominus_a (x^{-a} \odot y)$. So we get $y \in I$.

(2) $\implies$ (1) Let $x, y \in I$ and $a \in \mathbf{I}(A)$ with $x, y \leq a$. Then

$$x^{-a} \odot (x \ominus_a y) = x^{-a} \odot (x^{-a} \rightarrow_a y) \leq y.$$

Thus, we have $x^{-a} \odot (x \ominus_a y) \in I$ and so $x \ominus_a y \in I$ by (2). On the other hand, let $x \leq y$ and $y \in I$. For each $a \in \mathbf{I}(A)$ such that $x, y \leq a$, we have $y^{-a} \odot x \leq x^{-a} \odot x = 0$ and so $y^{-a} \odot x \in I$. Hence, we get $x \in I$.

(2) $\implies$ (3) Let $x, y \in A$ and $x, y \leq a$, where $a \in \mathbf{I}(A)$. Suppose that $(x^{-a} \rightarrow_a y^{-a})^{-a} \in I$ and $x \in I$. We have $x^{-a} \odot y^{-a-a} \leq (x^{-a} \odot y^{-a-a})^{-a-a} = (x^{-a} \rightarrow_a y^{-a})^{-a}$ and so $x^{-a} \odot y^{-a-a} \in I$. Thus $y^{-a-a} \in I$ by (2). From $y \leq y^{-a-a}$, it follows that $y \in I$.

(3) $\implies$ (2) For all $x, y \in A$ and $a \in \mathbf{I}(A)$ with $x, y \leq a$, suppose that $x^{-a} \odot y \in I$ and $x \in I$. We have $((x^{-a} \odot y)^{-a} \rightarrow_a ((x^{-a} \odot y)^{-a-a})^{-a})^{-a} = 0 \in I$. It follows that $(x^{-a} \odot y)^{-a-a} \in I$ by (3). From $(x^{-a} \rightarrow_a y^{-a})^{-a} = (x^{-a} \odot y)^{-a-a}$, we can obtain that $(x^{-a} \rightarrow_a y^{-a})^{-a} \in I$ and $y \in I$ using (3) again. $\square$

Remark 4.4. Let A be an Esemihoop with the least element 0 and I be an ideal of A. Then for all $x \in A$ and $a \in \mathbf{I}(A)$ such that $x \leq a$, we have $x \in I \iff x^{-a-a} \in I$. In fact, let $x \in I$ and $a \in \mathbf{I}(A)$ such that $x \leq a$, then $(x^{-a} \rightarrow_a x^{-a-a-a})^{-a} = 0 \in I$. By Proposition 4.3, $x^{-a-a} \in I$. The other direction is obvious.

Proposition 4.5. Let $f : A \rightarrow B$ be an Esemihoop homomorphism, where A and B are two Esemihoops with the least element 0. The following statements hold:
(1) if S is a subalgebra of A, then $f(S)$ is a subalgebra of B;
(2) if f is a bijection and I is an ideal of A, then $f(I)$ is an ideal of B;
(3) if f is surjective and I is an ideal of B, then $f^{-1}(I)$ is an ideal of A;
(4) $ker f = \{x \in A | f(x) = 0\}$ is an ideal of A.

169

Proof. (1) Suppose that S is a subalgebra of A. For $x, y \in S$, we have $f(x) \wedge f(y) = f(x \wedge y) \in f(S)$ and $f(x) \odot f(y) = f(x \odot y) \in f(S)$. For each $c \in \mathbf{I}(B) \cap f(S)$, there exists $a \in S$ such that $f(a) = c$. Let $f(S)_c := f(S) \cap B_c = \{z \in f(S) | z \leq c\}$. For all $x, y \in f(S)_c$, there are two elements $u, v \in S$ satisfying $f(u) = x$ and $f(v) = y$. Then $x \odot y = f(u) \odot f(v) = f(u \odot v) \in f(S)$. For $x, y \in f(S)_c$, we have $x \odot y \leq c$, which implies $x \odot y \in f(S)_c$. Similarly, $x \wedge y = f(u) \wedge f(v) = f(u \wedge v) \in f(S)_c$. It follows that $f(S)_c$ is closed under $\odot$ and $\wedge$. Moreover, since S is a subalgebra of A, there is $b \in \mathbf{I}(A) \cap S$ such that $u, v, a \leq b$. By Proposition 3.9, then

$$x \to_c y = f(u) \to_{f(a)} f(v) = (f(u) \to_{f(b)} f(v)) \wedge f(a) = f((u \to_b v) \wedge a) \in f(S)_c.$$

It follows that $f(S)_c$ is closed under $\to_c$. Hence, $f(S)_c$ is a subalgebra of the semi-hoop B_c.

For all $x, y \in f(S)$, there are $m, n \in S$ such that $f(m) = x$ and $f(n) = y$. We have an element $d \in \mathbf{I}(A) \cap S$ such that $m, n \leq d$. Then $x, y \leq f(d)$, where $f(d) \in \mathbf{I}(B) \cap f(S)$. This proves that $f(S)$ is a subalgebra of B.

(2) From $0 \in I$, we can get $0 = f(0) \in f(I)$. Suppose that $x \in f(I)$ and $x^{-b} \odot y \in f(I)$, where $b \in \mathbf{I}(B)$ and $x, y \leq b$. There exists $u \in I$ such that $x = f(u)$. Since f is bijective, we can find $w, a \in A$ such that $f(w) = y$ and $f(a) = b$. It follows that $f(u^{-a} \odot w) = f(u)^{-f(a)} \odot f(w) = x^{-b} \odot y \in f(I)$. Thus, $u^{-a} \odot w \in I$ and so $w \in I$. It follows that $y = f(w) \in f(I)$. This proves that $f(I)$ is an ideal of B.

By Proposition 4.3, the proofs of (3) and (4) are obvious. $\qquad \square$

Proposition 4.6. Let $(A, \wedge, \odot)$ be an Esemihoop with the least element 0. If θ is a congruence of A, then $0/\theta$ is an ideal of A.

Proof. Suppose that $x^{-a} \odot y \in 0/\theta$ and $x \in 0/\theta$, where $a \in \mathbf{I}(A)$ such that $x, y \leq a$. Then $(x^{-a} \odot y, 0) \in \theta$ and $(x, 0) \in \theta$. We have $(x^{-a} \odot y, 0^{-a} \odot y) \in \theta$. It follows that $(x^{-a} \odot y, y) \in \theta$. Thus, $(y, 0) \in \theta$ and so $y \in 0/\theta$. It is obvious that $0 \in 0/\theta$. Therefore, $0/\theta$ is an ideal by Proposition 4.3. $\qquad \square$

Proposition 4.7. Let A be a regular Esemihoop and I be an ideal of A. The relation θ_I defined by

$$(x, y) \in \theta_I \iff \exists \, a \in \mathbf{I}(A) \text{ such that } x, y \leq a \text{ and } (x \to_a y)^{-a}, (y \to_a x)^{-a} \in I$$

is a congruence on A.

Proof. It is easy to verify that θ_I is reflexive and symmetric. Assume that (x, y), $(y, z) \in \theta_I$. Then there exist $a, b \in \mathbf{I}(A)$ such that $x, y \leq a$, $y, z \leq b$ and $(x \to_a y)^{-a}, (y \to_a x)^{-a}, (y \to_b z)^{-b}, (z \to_b y)^{-b} \in I$. Let $c \in \mathbf{I}(A)$ such that $a, b \leq c$.

We have $(x \to_c y)^{-c} = x \odot y^{-c}$ and $(x \to_a y)^{-a} = x \odot y^{-a}$ by Proposition 3.12. Hence, $(x \to_c y)^{-c} = (x \to_a y)^{-a} \in I$ by Proposition 3.10. Similarly, we can obtain $(y \to_c x)^{-c}, (y \to_c z)^{-c}, (z \to_c y)^{-c} \in I$. From $z \to_c y \leq (y \to_c x) \to_c (z \to_c x)$, we have $((y \to_c x)^{-c-c} \to_c (z \to_c x)^{-c-c})^{-c} \leq (z \to_c y)^{-c} \in I$. Then $((y \to_c x)^{-c-c} \to_c (z \to_c x)^{-c-c})^{-c} \in I$ and so $(z \to_c x)^{-c} \in I$ by Proposition 4.3. In a similar way, $(x \to_c z)^{-c} \in I$. It follows that $(x, y) \in \theta_I$. Thus, θ_I is an equivalence relation. From Proposition 2.2(9), we easily get that θ_I is compatible with $\wedge$ and $\odot$.

Now, we show that $\theta_I \cap (A_a \times A_a)$ is a congruence on A_a for all $a \in \mathbf{I}(A)$. Assume $(x, y) \in \theta_I \cap (A_a \times A_a)$. Then there exists $b \in \mathbf{I}(A)$ such that $x, y \leq b$ and $(x \to_b y)^{-b}, (y \to_b x)^{-b} \in I$. Also, $(x \to_a y)^{-a}, (y \to_a x)^{-a} \in I$ by $(x, y) \in A_a \times A_a$. Let $z \in A_a$. From $x \to_a y \leq (y \to_a z) \to_a (x \to_a z)$, we get $((y \to_a z) \to_a (x \to_a z))^{-a} \leq (x \to_a y)^{-a} \in I$. Then $((y \to_a z) \to_a (x \to_a z))^{-a} \in I$. In a similar way, $((x \to_a z) \to_a (y \to_a z))^{-a} \in I$. Thus, $(x \to_a z, y \to_a z) \in \theta_I$. From $x \to_a y \leq (z \to_a x) \to_a (z \to_a y)$, we get $((z \to_a x) \to_a (z \to_a y))^{-a} \leq (x \to_a y)^{-a} \in I$. Hence, $((z \to_a x) \to_a (z \to_a y))^{-a} \in I$. Similarly, $((z \to_a y) \to_a (z \to_a x))^{-a} \in I$. Thus, $(z \to_a x, z \to_a y) \in \theta_I$. It follows that $\theta_I \cap (A_a \times A_a)$ is a congruence on A_a. Therefore, θ_I is a congruence on A. $\square$

Remark 4.8. From Proposition 3.10, 3.12 and Proposition 4.7, we can get

$$(x, y) \in \theta_I \iff \forall\, a \in \mathbf{I}(A) \text{ such that } x, y \leq a \text{ and } (x \to_a y)^{-a}, (y \to_a x)^{-a} \in I.$$

We denote the Esemihoop A/θ_I by A/I and call it the quotient Esemihoop algebra of A induced by I.

Proposition 4.9. Let A be a regular Esemihoop. Then there is a one-to-one correspondence between the set of all ideals of A and the set of all congruences on A.

Proof. Suppose that I is an ideal of A and θ_I is the congruence induced by I. We have $I = 0/\theta_I$. Let $x \in I$. For all $a \in \mathbf{I}(A)$ such that $x \leq a$, we have $(x \to_a 0)^{-a} = x^{-a-a} \in I$ and $(0 \to_a x)^{-a} = 0 \in I$. It follows that $(x, 0) \in \theta_I$, i.e. $x \in 0/\theta_I$. Conversely, if $x \in 0/\theta_I$, i.e. $(x, 0) \in \theta_I$, then for all $a \in \mathbf{I}(A)$ such that $x \leq a$, we have $(x \to_a 0)^{-a} \in I$. Therefore, we get $x \in I$.

Let θ be a congruence and $I = 0/\theta$. Then θ_I is the congruence induced by I. We claim $\theta = \theta_I$. If $(x, y) \in \theta_I$, there is $a \in \mathbf{I}(A)$ such that $x, y \leq a$ and $(x \to_a y)^{-a}, (y \to_a x)^{-a} \in I$. This implies that $((x \to_a y)^{-a}, 0) \in \theta$ and so $(x \to_a y, a) \in \theta$. Thus, $((x \to_a y) \to_a y, y) = ((x \to_a y) \to_a y, a \to_a y) \in \theta$. By Proposition 2.2(4), we have $(((x \to_a y) \to_a y) \wedge ((y \to_a x) \to_a x), y) = (((x \to_a y) \to_a y) \wedge ((y \to_a x) \to_a x), y \wedge ((y \to_a x) \to_a x)) \in \theta$. Similarly, we get $(((y \to_a x) \to_a x) \wedge ((x \to_a$

$y) \to_a y), x) \in \theta$. Therefore, $(x, y) \in \theta$. If $(x, y) \in \theta$ and $a \in \mathbf{I}(A)$ with $x, y \leq a$, we have $((x \to_a y)^{-a}, (y \to_a y)^{-a}) \in \theta$ and $((x \to_a x)^{-a}, (y \to_a x)^{-a}) \in \theta$. This imply that $(x \to_a y)^{-a}, (y \to_a x)^{-a} \in 0/\theta = I$. Thus, $(x, y) \in \theta_I$. $\square$

Definition 4.10. Let A be an Esemihoop and $\emptyset \neq F \subseteq A$. F is called a filter of A if it satisfies:
(ESF1) for all $x \in A$, there is $a \in \mathbf{I}(A) \cap F$ such that $x \leq a$;
(ESF2) $x \leq y$ and $x \in F \Longrightarrow y \in F$ for all $x, y \in A$;
(ESF3) for all $x, y \in F \Longrightarrow x \odot y \in F$.

A filter F is said to be proper if $F \neq A$. The following statement provides an equivalent notion of filters of Esemihoops.

Proposition 4.11. Let A be an Esemihoop and $\emptyset \neq F \subseteq A$. Then F is a filter of A if and only if
(ESF1) for all $x \in A$, there exists $a \in \mathbf{I}(A) \cap F$ such that $x \leq a$;
(ESF2$'$) $x \in F$ and $x \to_a y \in F \Longrightarrow y \in F$, where $a \in \mathbf{I}(A)$ such that $x, y \leq a$.

Proof. $\Longrightarrow$: Let $x \in F$ and $x \to_a y \in F$, where $a \in \mathbf{I}(A)$ such that $x, y \leq a$. Then $x \odot (x \to_a y) \in F$. From $x \odot (x \to_a y) \leq y$, we have $y \in F$.

$\Longleftarrow$: Let $x \in F$ and $x \leq y$. By (ESF1), there exists $a \in \mathbf{I}(A) \cap F$ such that $y \leq a$. It follows that $x \leq y \leq a$ and $a \in F$. Then we have $x \to_a y = a \in F$. Thus, $y \in F$ by (ESF2$'$). If $x, y \in F$ and $b \in \mathbf{I}(A)$ such that $x, y \leq b$, we get $x \to_b (y \to_b (x \odot y)) = (x \odot y) \to_b (x \odot y) = b \in F$. Then $y \to_b (x \odot y) \in F$ and so $x \odot y \in F$ by (ESF2$'$) again. Therefore, F is a filter of A. $\square$

Proposition 4.12. Let A be an Esemihoop and θ be a congruence of A. There exists $a \in \mathbf{I}(A)$ such that $a/\theta = \{x \in A | (a, x) \in \theta\}$ is a filter of A.

Proof. For all $x, y \in A$ and $a \in \mathbf{I}(A)$ such that $x, y \leq a$, we know $a \in \mathbf{I}(A) \cap a/\theta$. It follows that (ESF1) holds. Suppose that $x \in a/\theta$ and $x \to_a y \in a/\theta$. We have $(a, x) \in \theta$ and $(a, x \to_a y) \in \theta$. Then $(a \to_a y, x \to_a y) \in \theta$. It follows $(y, x \to_a y) \in \theta$. Thus, $(a, y) \in \theta$ and so $y \in a/\theta$. This shows that a/θ is a filter of A by Proposition 4.11. $\square$

Proposition 4.13. Let A be an Esemihoop and F be a filter of A. We define a binary relation θ_F on A as follows:

$$(x, y) \in \theta_F \Longleftrightarrow \exists\, a \in \mathbf{I}(A) \text{ such that } x, y \leq a,\ x \to_a y,\ y \to_a x \in F.$$

Then θ_F is a congruence on A.

Proof. It is easy to know that θ_F is an equivalence relation and is compatible with $\odot$ and $\wedge$ by Proposition 2.2(8) and (9). We cliam that $\theta_F \cap (A_a \times A_a)$ is a congruence on A_a for all $a \in \mathbf{I}(A)$. Suppose that $a \in \mathbf{I}(A)$ and $(x, y) \in \theta_F \cap (A_a \times A_a)$. There is $b \in \mathbf{I}(A)$ such that $x, y \le b$ and $x \to_b y, y \to_b x \in F$. Let $z \in A_a$. Choose $c \in \mathbf{I}(A)$ such that $a, b \le c$. We obtain that $x \to_c y, y \to_c x \in F$. Thus, in A_c, we have

$$
\begin{aligned}
(x \to_a z) \to_c (y \to_a z) &= (x \to_a z) \to_c ((y \to_c z) \wedge a) \\
&= ((x \to_a z) \to_c (y \to_c z)) \wedge ((x \to_a z) \to_c a) \text{ (Remark 3.2)} \\
&= ((x \to_a z) \to_c (y \to_c z)) \wedge c \\
&= (x \to_a z) \to_c (y \to_c z).
\end{aligned}
$$

From $y \to_c x \le (x \to_c z) \to_c (y \to_c z)$, we know $(x \to_c z) \to_c (y \to_c z) \in F$. By $(x \to_c z) \to_c (y \to_c z) \le (x \to_a z) \to_c (y \to_c z)$, we get $(x \to_a z) \to_c (y \to_c z) \in F$ and so $(x \to_a z) \to_c (y \to_a z) \in F$. Similarly, $(y \to_a z) \to_c (x \to_a z) \in F$. Hence, $(x \to_a z, y \to_a z) \in \theta_F$. Also $(z \to_a x, z \to_a y) \in \theta_F$. Therefore, $\theta_F \cap (A_a \times A_a)$ is a congruence on A_a for all $a \in \mathbf{I}(A)$. $\square$

Since θ_F is a congruence on the Esemihoop A, A/θ_F is a quotient Esemihoop algebra and denote A/θ_F by A/F.

Proposition 4.14. Let A be an Esemihoop with the least element 0 and F be a filter of A. Then the set $I_F = \{x \in A | \exists u \in F, \exists a \in \mathbf{I}(A) \text{ such that } x, u \le a \text{ and } x^{-a-a} \le u^{-a}\}$ is an ideal of A.

Proof. Let $x, y \in I_F$. There exist $u \in F$, $a \in \mathbf{I}(A)$ such that $x, u \le a$ and $x^{-a-a} \le u^{-a}$. Also, there are $v \in F$ and $b \in \mathbf{I}(A)$, which satisfies $y, v \le b$ and $y^{-b-b} \le v^{-b}$. Choose $c \in \mathbf{I}(A)$ with $a, b \le c$. It is easy to know that $u \odot v \in F$. From $x^{-a-a} \le u^{-a}$, we have $u \le u^{-a-a} \le x^{-a} \le x^{-c}$. Similarly, $v \le v^{-b-b} \le y^{-b} \le y^{-c}$ and so $y^{-c-c} \le v^{-c}$. Thus,

$$
\begin{aligned}
(x \ominus_c y)^{-c-c} = (x^{-c} \to_c y)^{-c-c} &\le (y^{-c} \to_c x^{-c-c})^{-c-c} \text{ (Proposition 2.3)} \\
&\le (x^{-c} \odot y^{-c})^{-c} = x^{-c} \to_c y^{-c-c} \\
&\le x^{-c} \to_c v^{-c} \le u \to_c v^{-c} \\
&= (u \odot v)^{-c}.
\end{aligned}
$$

It follows that $x \ominus_c y \in I_F$. Suppose that $x \le y$ and $y \in I_F$. There exist $v \in F$, $b \in \mathbf{I}(A)$ such that $y, v \le b$ and $y^{-b-b} \le v^{-b}$. From $x \le y$, we have $x^{-b-b} \le y^{-b-b} \le v^{-b}$. Thus, $x \in I_F$. Therefore, I_F is an ideal of A. $\square$

173

Remark 4.15. Let A be an Esemihoop with the least element 0 and F be a filter of A. The following conditions are equivalent:

(1) $x \in I_F$;

(2) $x \in J_F = \{x \in A | \exists u \in F, \exists a \in \mathbf{I}(A) \text{ such that } x, u \le a \text{ and } x \le u^{-a}\}$;

(3) $x \in L_F = \{x \in A | \exists a \in \mathbf{I}(A) \text{ such that } x \le a \text{ and } x^{-a} \in F\}$.

Proof. The proof is straightforward. $\square$

Proposition 4.16. Let A be an Esemihoop with the least element 0. For all $x, y, z \in A$:

(1) if $a \in \mathbf{I}(A)$ such that $x, y \in A_a$, we have $x, y \le x \ominus_a y$;

(2) if $x \le y$ and $a \in \mathbf{I}(A)$ such that $x, y, z \in A_a$, we have $x \ominus_a z \le y \ominus_a z$ and $z \ominus_a x \le z \ominus_a y$.

Proof. (1) Let $a \in \mathbf{I}(A)$ such that $x, y \in A_a$. From $y \odot x^{-a} \le y$, we have $y \le x^{-a} \to_a y = x \ominus_a y$. From $x \odot x^{-a} = 0 \le y$, we have $x \le x^{-a} \to_a y = x \ominus_a y$.

(2) Let $x \le y$ and $a \in \mathbf{I}(A)$ such that $x, y, z \in A_a$. From $x \le y$, we can get $z^{-a} \to_a x \le z^{-a} \to_a y$. That is, $z \ominus_a x \le z \ominus_a y$. Similarly, by $y^{-a} \le x^{-a}$, we have $x^{-a} \to_a z \le y^{-a} \to_a z$ and so $x \ominus_a z \le y \ominus_a z$. $\square$

Proposition 4.17. Let A be a regular Esemihoop. For all $a \in \mathbf{I}(A)$ such that $x, y, z, y_i \in A_a$, the following statements hold:

(1) $x \ominus_a y = y \ominus_a x$;

(2) $x \ominus_a (y \ominus_a z) = (x \ominus_a y) \ominus_a z$;

(3) if $\wedge_{i \in I} y_i$ exists, $x \ominus_a (\wedge_{i \in I} y_i) = \wedge_{i \in I} (x \ominus_a y_i)$;

(4) $x \wedge (y_1 \ominus_a y_2 \ominus_a \cdots \ominus_a y_n) \le (x \wedge y_1) \ominus_a (x \wedge y_2) \ominus_a \cdots \ominus_a (x \wedge y_n)$;

(5) if $a, b \in \mathbf{I}(A)$ with $x \le a \le b$, then $x^{-a} \ominus_b a^{-b} = x^{-b} \ominus_b a^{-b}$.

Proof. (1) It is obvious by Proposition 3.12.

(2) We have

$$
\begin{aligned}
x \ominus_a (y \ominus_a z) &= x^{-a} \to_a (y^{-a} \to_a z) \\
&= (x^{-a} \odot y^{-a}) \to_a z \\
&= (x^{-a} \to_a y)^{-a} \to_a z \\
&= (x \ominus_a y) \ominus_a z.
\end{aligned}
$$

(3) Suppose that $\wedge_{i \in I} y_i$ exists. From $\wedge_{i \in I} y_i \le y_i$ for each $i \in I$, we have $x \ominus_a (\wedge_{i \in I} y_i) \le x \ominus_a y_i$ by Proposition 4.16. It follows that $x \ominus_a (\wedge_{i \in I} y_i) \le \wedge_{i \in I} (x \ominus_a y_i)$. Let $u = \wedge_{i \in I} (x \ominus_a y_i)$. We have $u = \wedge_{i \in I} (x \ominus_a y_i) \le x \ominus_a y_i = y_i \ominus_a x = y_i^{-a} \to_a x$ for all $i \in I$. Then $y_i^{-a} \le u \to_a x$. It follows that $(u \to_a x)^{-a} \le y_i^{-a-a} = y_i$ and so $(u \to_a x)^{-a} \le \wedge_{i \in I} y_i$. Thus, $(\wedge_{i \in I} y_i)^{-a} \le (u \to_a x)^{-a-a} = u \to_a x$. Hence,

$u \leq (\wedge_{i \in I} y_i)^{-a} \to_a x = (\wedge_{i \in I} y_i) \ominus_a x = x \ominus_a (\wedge_{i \in I} y_i)$. So $\wedge_{i \in I}(x \ominus_a y_i) \leq x \ominus_a (\wedge_{i \in I} y_i)$.
Therefore, $x \ominus_a (\wedge_{i \in I} y_i) = \wedge_{i \in I}(x \ominus_a y_i)$.

(4) Firstly, for $n = 2$, we have

$$
\begin{aligned}
(x \wedge y_1) \ominus_a (x \wedge y_2) &= ((x \wedge y_1) \ominus_a x) \wedge ((x \wedge y_1) \ominus_a y_2) \\
&= (x \ominus_a x) \wedge (x \ominus_a y_1) \wedge (y_2 \ominus_a x) \wedge (y_2 \ominus_a y_1) \\
&\geq x \wedge x \wedge x \wedge (y_1 \ominus_a y_2) \\
&= x \wedge (y_1 \ominus_a y_2).
\end{aligned}
$$

By inductive hypothesis, we can get $x \wedge (y_1 \ominus_a y_2 \ominus_a \cdots \ominus_a y_n) \leq (x \wedge y_1) \ominus_a (x \wedge y_2) \ominus_a \cdots \ominus_a (x \wedge y_n)$.

(5) From Proposition 3.9(1) and Proposition 4.17(3), we have

$$
\begin{aligned}
x^{-a} \ominus_b a^{-b} &= (x^{-b} \wedge a) \ominus_b a^{-b} \\
&= (x^{-b} \ominus_b a^{-b}) \wedge (a \ominus_b a^{-b}) \\
&= x^{-b} \ominus_b a^{-b}.
\end{aligned}
$$

$\square$

Proposition 4.18. Let A be a regular Esemihoop and I a proper ideal of A. If I satisfies the condition:

$$\text{for all } a \in \mathbf{I}(A) \text{ and } a \notin I \implies \text{ for all } b \in \mathbf{I}(A)$$
$$\text{such that } a \leq b, \text{ we have } a^{-b} \in I. \ (*)$$

Then $F_I = \{x \in A | \exists a \in \mathbf{I}(A) \backslash I \text{ such that } x \leq a \text{ and } x^{-a} \in I\}$ is a filter of A.

Proof. Since I is a proper ideal, there exists $x \in A \backslash I$. Let $a \in \mathbf{I}(A)$ such that $x \leq a$. We have $a \notin I$. For all $y \in A$, there exists $b \in \mathbf{I}(A)$ such that $y \leq b$. Assume that $c \in \mathbf{I}(A)$ with $a, b \leq c$. We have $c \notin I$. From $(*)$, we have $c^{-c} \in I$ and so $c \in F_I$. This means that $F_I \neq \emptyset$ and for all $y \in A$, there is $c \in \mathbf{I}(A) \cap F_I$ such that $y \leq c$.

Suppose that $x \leq y$ and $x \in F_I$. There is $a \in \mathbf{I}(A) \backslash I$ such that $x \leq a$ and $x^{-a} \in I$. For all $b \in \mathbf{I}(A)$ such that $x, y, a \leq b$, we get $a^{-b} \in I$ by $(*)$ and $b \notin I$. By Proposition 4.17(5) and Proposition 4.16(1), we have $x^{-b} \leq x^{-b} \ominus_b a^{-b} = x^{-a} \ominus_b a^{-b} \in I$. It follows that $x^{-b} \in I$. From $y^{-b} \leq x^{-b}$, we get $y^{-b} \in I$. This means that $y \in F_I$.

Suppose $x, y \in F_I$. There are $a, b \in \mathbf{I}(A) \backslash I$ such that $x \leq a$, $y \leq b$ and $x^{-a}, y^{-b} \in I$. Let $c \in \mathbf{I}(A)$ such that $a, b \leq c$. We have $c \notin I$. It follows that $a^{-c}, b^{-c} \in I$ and $c \in \mathbf{I}(A) \backslash I$. Similarly, from $x^{-c} \leq x^{-c} \ominus_c a^{-c} = x^{-a} \ominus_c a^{-c} \in I$ and $y^{-c} \leq y^{-c} \ominus_c b^{-c} = y^{-b} \ominus_c b^{-c} \in I$, we get $x^{-c}, y^{-c} \in I$. Then, $(x \odot y)^{-c} = x \to_c y^{-c} = x^{-c-c} \to_c y^{-c} = x^{-c} \ominus_c y^{-c} \in I$, which implies that $x \odot y \in F_I$. Therefore, F_I is a filter of A. $\square$

175

5 Maximal filters, maximal ideals and prime ideals

In this section, we shall introduce maximal filters, maximal ideals and prime ideals of Esemihoops and give a prime ideal theorem. We prove that every Esemihoop has at least one maximal filter. If A is a regular Esemihoop, A has at least one maximal ideal and every maximal ideal is prime. In addition, it is proved that every proper ideal of A is contained in a maximal ideal.

Let A be an Esemihoop. If $F \subseteq A$, the intersection of all filters containing F is denoted by $\lfloor F \rceil$, which is called the filter generated by F. If $F = \{x\}$, we write $\lfloor x \rceil$ as alternatives to $\lfloor \{x\} \rceil$.

For all integer $n \geq 2$ and $x \in A$, we define $x^n = x^{n-1} \odot x$.

Proposition 5.1. Let F be a filter of an Esemihoop A. Then for all $x \in A$,
(1) $\lfloor x \rceil = \{u \in A | u \geq x^m, \text{ for some } m \in \mathbb{N}^*\}$, where $\mathbb{N}^* = \{1, 2, 3, 4, 5, \cdots\}$;
(2) $\lfloor F \cup \{x\} \rceil = \{u \in A | u \geq y \odot x^m, \text{ for some } y \in F \text{ and } m \in \mathbb{N}^*\}$.

Proof. (1) Suppose $S = \{u \in A | u \geq x^m, \text{ for some } m \in \mathbb{N}^*\}$. Obviously, $x \in S$. Thus, $S \neq \emptyset$. For all $y \in A$, there is $a \in \mathbf{I}(A)$ such that $x, y \leq a$. We have $a \in S$ and so $a \in \mathbf{I}(A) \cap S$. It follows that (ESF1) holds. Let $u, u \rightarrow_b v \in S$ and $b \in \mathbf{I}(A)$ such that $u, v \leq b$. There are $m, n \in \mathbb{N}^*$ such that $u \geq x^m$ and $u \rightarrow_b v \geq x^n$. Then $x^{m+n} = x^m \odot x^n \leq u \odot (u \rightarrow_b v) \leq v$, which implies that $v \in S$. Hence, we have shown that S is a filter containing $\{x\}$. Suppose that T is a filter containing $\{x\}$. For all $u \in S$, there exists $m \in \mathbb{N}^*$ such that $u \geq x^m$. As $x^m \in T$, we have $u \in T$. Thus, $S \subseteq T$. Therefore, $\lfloor x \rceil = \{u \in A | u \geq x^m, \text{ for some } m \in \mathbb{N}^*\}$.

(2) The proof is similar to (1). □

Definition 5.2. Let A be an Esemihoop. A proper filter F of A is maximal if no other proper filters of A can strictly contain F.

Theorem 5.3. Let A be an Esemihoop and F be a filter of A. The following statements are equivalent:
(1) F is a maximal filter;
(2) if $x \in A \backslash F$, then $\lfloor F \cup \{x\} \rceil = A$;
(3) if $x \in A \backslash F$, then for all $u \in A$, we have $x^m \rightarrow_a u \in F$, where $m \in \mathbb{N}^*$ and $a \in \mathbf{I}(A)$ with $x, u \leq a$.

Proof. (1) $\Longrightarrow$ (2) Suppose that F is a maximal filter. If $x \in A \backslash F$, then $F \subsetneq F \cup \{x\} \subseteq \lfloor F \cup \{x\} \rceil$. Thus, $\lfloor F \cup \{x\} \rceil = A$.

(2) $\Longrightarrow$ (1) If H is a filter such that $F \subsetneq H$, there exists $x \in H \backslash F$. It follows that $\lfloor F \cup \{x\} \rceil = A$ by (2). From $F \subseteq H$ and $x \in H$, we have $F \cup \{x\} \subseteq H$ and so

176

$\lfloor F \cup \{x\} \rfloor \subseteq H$. Hence, $A \subseteq H$, which implies $H = A$. Therefore, F is a maximal filter.

(2) $\Longrightarrow$ (3) Let $\lfloor F \cup \{x\} \rfloor = A$, where $x \in A \backslash F$. For all $u \in A$, there exist $y \in F$ and $m \in \mathbb{N}^*$ such that $u \geq y \odot x^m$. Take $a \in \mathbf{I}(A)$ such that $x, y, u \leq a$. In the semihoop A_a, we have $y \leq x^m \rightarrow_a u$. It means that $x^m \rightarrow_a u \in F$.

(3) $\Longrightarrow$ (2) Suppose $x \in A \backslash F$. For all $u \in A$, there exist $m \in \mathbb{N}^*$ and $a \in \mathbf{I}(A)$ such that $x, u \leq a$ and $x^m \rightarrow_a u \in F$. Since $x^m \odot (x^m \rightarrow_a u) \leq u$, we can get $u \in \lfloor F \cup \{x\} \rfloor$ and so $A \subseteq \lfloor F \cup \{x\} \rfloor$. Therefore, $\lfloor F \cup \{x\} \rfloor = A$. $\square$

Proposition 5.4. Let F be a maximal filter of an Esemihoop A with the least element 0. Then $x \in A \backslash F \Longleftrightarrow \exists a \in \mathbf{I}(A), \exists m \in \mathbb{N}^*$ such that $(x^m)^{-a} \in F$.

Proof. By Proposition 5.1 and Theorem 5.3, the proof is obvious. $\square$

Proposition 5.5. Let F be a maximal filter of an Esemihoop A. Then for all $a \in \mathbf{I}(A)$, $F \cap A_a = \emptyset$ or $F \cap A_a$ is a maximal filter of the semihoop $(A_a, \odot, \rightarrow_a, \wedge, 0, a)$.

Proof. Suppose $a \in \mathbf{I}(A)$ and $F \cap A_a \neq \emptyset$. There exists $x \in F \cap A_a$, which means that $a \in F$. Obviously, $F \cap A_a$ is a filter of A_a. For all $x \in A_a \backslash (F \cap A_a)$, we have $\lfloor F \cup \{x\} \rfloor = A$ by Proposition 5.3. For arbitrary $u \in A_a \subseteq \lfloor F \cup \{x\} \rfloor$, there exist $y \in F$ and $m \in \mathbb{N}^*$ such that $u \geq y \odot x^m$. Let $b \in \mathbf{I}(A)$ such that $a, y \leq b$. We have $u = u \wedge a \geq (y \odot x^m) \wedge a = (y \odot x^m) \odot a = (y \odot a) \odot x^m$. From $y \odot a \leq y \wedge a \leq a$ and $y, a \in F$, we have $y \odot a \in F \cap A_a$. It follows that $u \in \lfloor (F \cap A_a) \cup \{x\} \rfloor$. Hence $A_a = \lfloor (F \cap A_a) \cup \{x\} \rfloor$. This shows that $F \cap A_a$ is a maximal filter. $\square$

Proposition 5.6. Let A be an Esemihoop with the least element 0 and F be a proper filter of A. The following statements hold:
(1) if $x \in F$, then $x \notin I_F$;
(2) if $x \in F$, for any $a \in \mathbf{I}(A)$ with $x \leq a$, we have $x^{-a} \in I_F$;
(3) if F is a maximal filter, for any $a \in \mathbf{I}(A)$ such that $a \notin I_F$, we have $a \in F$.

Proof. (1) Suppose that $x \in I_F$ for any $x \in F$. There exist $u \in F$ and $a \in \mathbf{I}(A)$ such that $x, u \leq a$ and $x \leq u^{-a}$ by Remark 4.15. By $x \in F$, we get $u^{-a} \in F$. Hence, $0 = u \odot u^{-a} \in F$, which is a contradiction.

(2) It is obvious.

(3) If $a \notin F$, we have $\lfloor F \cup \{a\} \rfloor = A$. From $0 \in A$, there exist $u \in F$ and $m \in \mathbb{N}^*$ such that $0 \geq u \odot a^m$. Take $b \in \mathbf{I}(A)$ such that $u, a \leq b$. We have $a^m \leq u \rightarrow_b 0 = u^{-b}$. Thus, $a = a^m \leq u^{-b}$ and so $a \in I_F$, which is a contradiction. $\square$

Proposition 5.7. Let F be a maximal filter of an Esemihoop A. Then A/F is a simple semihoop.

Proof. Since F is a maximal filter, for all $x \in A$, there exists $a \in \mathbf{I}(A) \cap F$ such that $x \leq a$. We have $x/F \leq a/F = F$, which implies that F is the top element of the Esemihoop A/F. Thus, A/F is a semihoop. Suppose that $\mathcal{H}$ is a proper filter of A/F. Clearly, $F \in \mathcal{H}$. Set $\hat{F} = \{x \in A | x/F \in \mathcal{H}\}$. We have that $F \subseteq \hat{F}$ and $\hat{F}$ is a filter of A/F. If there exists $x \in \hat{F} \backslash F$, then we have $x/F \in \mathcal{H}$ and $x \notin F$. Take $y \in A$ such that $y/F \notin \mathcal{H}$. From Theorem 5.3, there exist $m \in \mathbb{N}^*$ and $b \in \mathbf{I}(A)$ such that $x, y \leq b$ and $x^m \to_b y \in F$. It follows that $b \in F$. Thus, we have $(x^m \to_b y)/F = F = b/F$ and so $x^m/F \to_{b/F} y/F = b/F$. Then $x^m/F \leq y/F$. This means that $x^m/F \notin \mathcal{H}$ and so $x^m \notin \hat{F}$, which is a contradiction. We have $\hat{F} = F$. Thus, $\mathcal{H} = \{F\}$, which implies that A/F has only two filters. $\square$

Remark 5.8. Let A be a regular Esemihoop. For all $x \in A$ and $a \in \mathbf{I}(A)$ such that $x \in A_a$, we define $2_a x = x \ominus_a x$, $3_a x = x \ominus_a 2_a x$, $\cdots$, $n_a x = x \ominus_a (n-1)_a x$. By Proposition 4.16, it is easy to know that $2_a x \leq 3_a x \leq \cdots n_a x \leq \cdots$. That is, taking $m, n \in \mathbb{N}^*$ and $m \leq n$, we can get $m_a x \leq n_a x$. If there is $b \in \mathbf{I}(A)$ such that $a \leq b$, by Proposition 3.9(4), we have $n_a x \leq n_b x$. Moreover, for all $n \in \mathbb{N}^*$ and $x_i \in A_a (i = 1, \cdots, n)$, define $\ominus_{a_{i=1}}^n x_i = x_1 \ominus_a x_2 \ominus_a \cdots \ominus_a x_n$.

For all subset X of A, the intersection of all ideals containing X is an ideal. We denote it by $\langle X \rangle$ and call it the ideal of A generated by X. If $X = \{x\}$, we write $\langle x \rangle$ as alternatives to $\langle \{x\} \rangle$.

Proposition 5.9. Let A be a regular Esemihoop and I be an ideal of A. Then for all $x \in A$, we have
(1) $\langle x \rangle = \{u \in A | u \leq m_a x$, for some $m \in \mathbb{N}^*$ and $a \in \mathbf{I}(A)$ such that $x \leq a\}$;
(2) $\langle I \cup \{x\} \rangle = \{u \in A | u \leq \ominus_{a_{i=1}}^k (\alpha_i \ominus_a m_{ia} x)$, for some $k, m_i \in \mathbb{N}^*$ and some $\alpha_i \in I, a \in \mathbf{I}(A)$ such that $\alpha_i, x \leq a\}$.

Proof. (1) By Definition 4.2 and Proposition 4.16, the proof is obvious.
   (2) Let $G = \{u \in A | u \leq \ominus_{a_{i=1}}^k (\alpha_i \ominus_a m_{ia} x)$, for some $k, m_i \in \mathbb{N}^*$ and some $\alpha_i \in I, a \in \mathbf{I}(A)$ such that $\alpha_i, x \leq a\}$. It is easy to know that $0 \in G$ and $I \cup \{x\} \subseteq G$. Let $u^{-a} \odot v \in G$ and $u \in G$, where $a \in \mathbf{I}(A)$ such that $u, v \leq a$. There are $\alpha_i \in I$, $k, m_i \in \mathbb{N}^*$ and $b \in \mathbf{I}(A)$ such that $x, \alpha_i \leq b$ and $u^{-a} \odot v \leq \ominus_{b_{i=1}}^k (\alpha_i \ominus_b m_{ib} x)$. Also, we have $\beta_i \in I, t, n_i \in \mathbb{N}^*$ and $c \in \mathbf{I}(A)$ such that $x, \beta_i \leq c$ and $u \leq \oplus_{c_{i=1}}^t (\beta_i \ominus_c n_{ic} x)$. Suppose $d \in \mathbf{I}(A)$ such that $a, b, c \leq d$, we have

$$v \leq u^{-d} \to_d (u^{-d} \odot v) = u^{-d} \to_d (u^{-a} \odot v)$$
$$\leq (\ominus_{c_{i=1}}^t (\beta_i \ominus_c n_{ic} x))^{-d} \to_d (\ominus_{b_{i=1}}^k (\alpha_i \ominus_b m_{ib} x))$$
$$= (\ominus_{c_{i=1}}^t (\beta_i \ominus_c n_{ic} x)) \ominus_d (\ominus_{b_{i=1}}^k (\alpha_i \ominus_b m_{ib} x))$$
$$\leq (\ominus_{d_{i=1}}^t (\beta_i \ominus_d n_{id} x)) \ominus_d (\ominus_{d_{i=1}}^k (\alpha_i \ominus_d m_{id} x)).$$

It follows that $v \in G$. Thus, G is an ideal of A containing $I \cup \{x\}$.

Suppose that J is an ideal of A containing $I \cup \{x\}$. For all $u \in G$, there are $\beta_i \in I$, $t, n_i \in \mathbb{N}^*$ and $c \in \mathbf{I}(A)$ such that $x, \beta_i \leq c$ and $u \leq \ominus_{c_{i=1}}^{t}(\beta_i \ominus_c n_{ic}x)$. Since $I \cup \{x\} \subseteq J$, we have $\beta_i \ominus_c n_{ic}x \in J$ and so $u \leq \ominus_{c_{i=1}}^{t}(\beta_i \ominus_c n_{ic}x) \in J$. It follows that $u \in J$, which implies that $G \subseteq J$. Therefore, G is the ideal of A generated by $I \cup \{x\}$. $\qquad \square$

Definition 5.10. Let I be a proper ideal of an Esemihoop A. I is maximal if it is not properly contained in any other proper ideal of A.

If A is a regular Esemihoop, then I is a maximal ideal if and only if for all $x \in A \backslash I$, $\langle I \cup \{x\} \rangle = A$.

Proposition 5.11. Let I be a maximal filter of a regular Esemihoop A. Then $I_F = \{x \in A | \exists f \in F, \exists a \in \mathbf{I}(A) \text{ such that } x, f \leq a \text{ and } x \leq f^{-a}\}$ is a maximal ideal of A.

Proof. By Proposition 4.14 and Remark 4.15, I_F is an ideal of A. Since F is a maximal filter, for all $x \in A$, there is $a \in \mathbf{I}(A) \cap F$ such that $x \leq a$. From $a \in F$ and Proposition 5.6(1), we have $a \notin I_F$. Hence, I_F is a proper ideal of A.

Assume that J is a proper ideal of A and $I_F \subseteq J$. If $a \in \mathbf{I}(A)$ and $a \notin J$, we have $a \notin I_F$. By Proposition 5.6(3), we get $a \in F$. Then for all $b \in \mathbf{I}(A)$ such that $a \leq b$, we obtain $a^{-b} \in I_F \subseteq J$. It follows from Proposition 4.18 that F_J is a filter of A. Suppose $h \in A \backslash J$. For all $f \in F$, there exists $c \in \mathbf{I}(A)$ such that $f, h \leq c$. Hence, we have $c \notin J$ and $f^{-c} \in I_F \subseteq J$, which implies that $f \in F_J$ and so $F \subseteq F_J$. From the maximality of F, $F_J = F$ or $F_J = A$. If $F_J = A$, we have $0 \in F_J$. There exists $d \in \mathbf{I}(A) \backslash J$ such that $d = 0^{-d} \in J$, which is a contradiction. Thus, $F_J = F$.

Let $x \in J$. There exists $a \in \mathbf{I}(A) \backslash J$ such that $x \leq a$. From $x^{-a-a} = x \in J$, we have $x^{-a} \in F_J = F$. It follows from $x \leq x^{-a-a}$ that $x \in I_F$. This proves that $J \subseteq I_F$ and so $J = I_F$. Thus, I_F is a maximal ideal of A. $\qquad \square$

Theorem 5.12. Let A be a proper Esemihoop. A has at least one maximal filter. If A is regular, then A has at least one maximal ideal.

Proof. Suppose that $\mathcal{F}$ is the set of all proper filters of A. There exists $0 \neq a \in \mathbf{I}(A)$. It is easy to know that the set $\{x \in A | x \geq a\}$ is a proper filter, which implies that $\mathcal{F} \neq \emptyset$. By Zorn's Lemma, there exists a maximal element F in $\mathcal{F}$. Therefore, A has at least one maximal filter.

If A is regular, by Proposition 5.11, we have that I_F is a maximal ideal of A. $\quad \square$

Definition 5.13. Let P be a proper ideal of an Esemihoop A. P is said to be a prime ideal of A if $x \wedge y \in P$ implies that $x \in P$ or $y \in P$ for all $x, y \in A$.

Theorem 5.14. (Prime ideal theorem) Let A be a regular Esemihoop and I an ideal of A. Suppose $\emptyset \neq S \subseteq A$ such that $I \cap S = \emptyset$. If S is closed under $\wedge$, there exists a prime ideal P of A such that $I \subseteq P$ and $P \cap S = \emptyset$.

Proof. Let $H = \{J$ is an ideal of $A | I \subseteq J$ and $J \cap S = \emptyset\}$. Obviously, we have $I \in H$ and $H \neq \emptyset$. By Zorn's Lemma, there exists a maximal element P of H. We have that $I \subseteq P$ and $P \cap S = \emptyset$. It follows that $P \neq A$. Suppose that $x \wedge y \in P$. If $x \notin P$ and $y \notin P$, then we have $P \subsetneq \langle P \cup \{x\} \rangle$ and $P \subsetneq \langle P \cup \{y\} \rangle$. Since P is maximal, we obtain that $S \cap \langle P \cup \{x\} \rangle \neq \emptyset$ and $S \cap \langle P \cup \{y\} \rangle \neq \emptyset$. We have $u \in S$ such that $u \in \langle P \cup \{x\} \rangle$. There are $\alpha_i \in P$, $k, m_i \in \mathbb{N}^*$ and $a \in \mathbf{I}(A)$ such that $\alpha_i, x \leq a$ and $u \leq \ominus_{a_{i=1}}^{k}(\alpha_i \ominus_a m_{ia}x)$. Similarly, we have $v \in S$ and $v \leq \ominus_{b_{i=1}}^{t}(\beta_i \ominus_b n_{ib}y)$, where $\beta_i \in P$, $t, n_i \in \mathbb{N}^*$ and $b \in \mathbf{I}(A)$ such that $\beta_i, y \leq b$. Let $c \in \mathbf{I}(A)$ with $a, b \leq c$, $z = (\ominus_{c_{i=1}}^{k}\alpha_i) \ominus_c (\ominus_{c_{i=1}}^{t}\beta_i) \in P$ and $n = max\{m_1, m_2, \cdots, m_k, n_1, n_2, \cdots, n_t\}$. By Proposition 3.9 and 4.17 and Remark 5.8, we can get

$$
\begin{aligned}
u \wedge v &\leq (\ominus_{a_{i=1}}^{k}(\alpha_i \ominus_a m_{ia}x)) \wedge (\ominus_{b_{i=1}}^{t}(\beta_i \ominus_b n_{ib}y)) \\
&\leq (\ominus_{c_{i=1}}^{k}(\alpha_i \ominus_c m_{ic}x)) \wedge (\ominus_{c_{i=1}}^{t}(\beta_i \ominus_c n_{ic}y)) \\
&\leq k_c(z \ominus_c n_c x) \wedge t_c(z \ominus_c n_c y) \\
&\leq (kt)_c((z \ominus_c n_c x) \wedge (z \ominus_c n_c y)) \\
&= (kt)_c(z \ominus_c (n_c x \wedge n_c y)) \\
&\leq (kt)_c(z \ominus_c n_c^2(x \wedge y)).
\end{aligned}
$$

It follows from $x \wedge y \in P$ and $z \in P$ that $u \wedge v \in P$. Since S is closed under $\wedge$, we know that $u \wedge v \in S$ and so $u \wedge v \in P \cap S$, which is a contradiction. Thus, P is a prime ideal of A. $\qquad\square$

Corollary 5.15. Let A be a regular Esemihoop. If I is an ideal of A and $x \in A \backslash I$, there exists a prime ideal P of A such that $I \subseteq P$ and $x \notin P$.

By Corollary 5.15, the following statement is direct.

Proposition 5.16. Let A be a regular Esemihoop. Then every maximal ideal I of A is a prime ideal.

Proposition 5.17. Let P be a prime ideal of a regular Esemihoop A. Then P is contained in a maximal ideal of A.

Proof. Assume that $x \in P$ and $a \in \mathbf{I}(A) \backslash P$ such that $x \leq a$. We have $x^{-a-a} \in P$ by Remark 4.4. It follows that $x^{-a} \in F_P$ and $x^{-a-a} \in I_{F_P}$. From $x \leq x^{-a-a}$, we have $x \in I_{F_P}$. Thus, $P \subseteq I_{F_P}$. Let $a \in \mathbf{I}(A)$ and $a \notin P$. For all $b \in \mathbf{I}(A)$ such that $a \leq b$, we have $a \wedge a^{-b} = a^{-a} = 0 \in P$. Since P is prime and $a \notin P$, we get $a^{-b} \in P$. This proves that P satisfies the condition in Proposition 4.18. Hence, F_P is a filter of A. By Proposition 4.14 and Remark 4.15, I_{F_P} is an ideal.

Set $\mathcal{F} = \{F | F$ is a filter of A and $F_P \subseteq F\}$. By Zorn's Lemma, there is a maximal element F_{max} in $\mathcal{F}$. This implies that F_{max} is a maximal filter of A and $F_P \subseteq F_{max}$. Thus, $I_{F_P} \subseteq I_{F_{max}}$. It follows that $P \subseteq I_{F_{max}}$. Since F_{max} is maximal, we have that $I_{F_{max}}$ is a maximal ideal by Proposition 5.11. $\qquad\square$

By Corollary 5.15 and Proposition 5.17, we get the following statement.

Proposition 5.18. *Every proper ideal of a regular Esemihoop A is contained in a maximal ideal of A.*

Proposition 5.19. *Let P is an ideal of a regular Esemihoop A. Then P is prime if and only if for any two ideals I and J, $I \cap J \subseteq P$ implies that $I \subseteq P$ or $J \subseteq P$.*

Proof. $\Longrightarrow$: Suppose that $I \nsubseteq P$ and $J \nsubseteq P$. There are $x \in I \backslash P$ and $y \in J \backslash P$. It follows that $x \wedge y \in I \cap J \subseteq P$. Since P is prime, we have $x \in P$ or $y \in P$, which is a contradiction. Thus, $I \subseteq P$ or $J \subseteq P$.

$\Longleftarrow$: For all $x, y \in A$, we have $\langle x \wedge y \rangle = \langle x \rangle \cap \langle y \rangle$. Indeed, Let $z \in \langle x \wedge y \rangle$. There exist $a \in \mathbf{I}(A)$ and $n \in \mathbb{N}^*$ such that $x \wedge y \leq a$ and $z \leq n_a(x \wedge y)$. For all $c \in \mathbf{I}(A)$ such that $x, y, a \leq c$, by Proposition 4.16 and Remark 5.8, we have $n_a(x \wedge y) \leq n_a x \leq n_c x$ and $n_a(x \wedge y) \leq n_a y \leq n_c y$. Thus, $z \leq n_c x$ and $z \leq n_c y$, which imply that $z \in \langle x \rangle \cap \langle y \rangle$. Conversely, let $z \in \langle x \rangle \cap \langle y \rangle$. There are $a, b \in \mathbf{I}(A)$ and $m, n \in \mathbb{N}^*$ such that $x \leq a, y \leq b, z \leq m_a x$ and $z \leq n_b y$. Then, for all $c \in \mathbf{I}(A)$ with $a, b \leq c$, we have $z \leq m_a x \wedge n_b y \leq m_c x \wedge n_c y \leq (mn)_c(x \wedge y)$ by Proposition 4.17. Thus, $z \in \langle x \wedge y \rangle$. If $x \wedge y \in P$, we get $\langle x \rangle \cap \langle y \rangle = \langle x \wedge y \rangle \subseteq P$. Hence, $\langle x \rangle \subseteq P$ or $\langle y \rangle \subseteq P$. This proves that $x \in P$ or $y \in P$. Therefore, P is prime. $\qquad\square$

Proposition 5.20. *Let A be a regular Esemihoop and I a maximal ideal of A. For all $a \in \mathbf{I}(A)$, $I \cap A_a = A_a$ or $I \cap A_a$ is a maximal ideal of the semihoop $(A_a, \odot, \rightarrow_a, \wedge, 0, a)$.*

Proof. Assume that $a \in \mathbf{I}(A)$ and $I \cap A_a \neq A_a$. Obviously, $I \cap A_a$ is an ideal of the semihoop A_a. Let $x \in A_a \backslash (I \cap A_a)$. We have $x \notin I$ and so $\langle I \cup \{x\} \rangle = A$. For all $z \in A_a \subseteq A$, there exist $\omega_i \in I, s, n_i \in \mathbb{N}^*$ and $b \in \mathbf{I}(A)$ such that $\omega_i, x \leq b$ and $z \leq \ominus_{b_{i=1}}^s (\omega_i \ominus_b n_{ib} x)$. Hence, there is $c \in \mathbf{I}(A)$ such that $a, b \leq c$, we obtain

181

$z \leq \ominus_{c i=1}^{s}(\omega_i \ominus_c n_{ic}x)$. By Proposition 4.17, we have

$$\begin{aligned}
z = z \wedge a &\leq (\ominus_{c i=1}^{s}(\omega_i \ominus_c n_{ic}x)) \wedge a \\
&\leq \ominus_{c i=1}^{s}((\omega_i \ominus_c n_{ic}x) \wedge a) \\
&\leq \ominus_{c i=1}^{s}((\omega_i \wedge a) \ominus_c (n_{ic}x \wedge a)) \\
&\leq \ominus_{c i=1}^{s}((\omega_i \wedge a) \ominus_c n_{ic}x).
\end{aligned}$$

Since $\omega_i \wedge a \in I \cap A_a$, we get $z \in \langle (I \cap A_a) \cup \{x\} \rangle$. This implies that $\langle (I \cap A_a) \cup \{x\} \rangle = A_a$. Therefore, $I \cap A_a$ is a maximal ideal of A_a. $\qquad\square$

6   States

In this section, we shall define Bosbach states and Riečan states on Esemihoops. In addition, the relations between Bosbach states and Riečan states are discussed. It is proved that a Bosbach state on Esemihoops is a Riečan state. In particular, Bosbach states and Riečan states are consistent on Glivenko Esemihoops.

Definition 6.1. Let A be an Esemihoop with the least element 0. A Bosbach state on A is a function $s : A \to [0,1]$ satisfying the following conditions:
(1) $s(0) = 0$;
(2) there exists $x_0 \in A$ such that $s(x_0) = 1$;
(3) for all $x, y \in A$ and $a \in \mathbf{I}(A)$ such that $x, y \leq a$, $s(x) + s(x \to_a y) = s(y) + s(y \to_a x)$.

Example 6.2. For the bounded semihoop $A = \{0, a, b, c, 1\}$ given in Example 3.15(ii) of [2], let $B = \{f \in \prod_{i \in I} A_i | supp(f)$ is finite$\}$, where $A_i = A$ for each $i \in I$. From Example 3.5, we have that B is an Esemihoop. For all $f \in B$, we define a mapping s on B. If for each $i \in I$, $f_i = 0$, then $s(f) = 0$. Otherwise, $s(f) = 1$. For all $f, g \in B$, there exists $m \in \mathbf{I}(B)$ such that $f, g \leq m$. It is easy to check that $s(f) + s(f \to_m g) = s(g) + s(g \to_m f)$. Therefore, s is a Bosbach state on B.

Example 6.3. Let $A = \{0, a, b, 1\}$ be the bounded perfect semihoop in Example 4.6(ii) of [2]. By Theorem 3.5 in [3], we have that A has a Bosbach state s. For all $x \in A$,

$$s(x) = \begin{cases} 0, & x = 0, \\ 1, & otherwise. \end{cases}$$

Suppose that $B = \{f \in \prod_{i \in I} A_i | supp(f)$ is finite$\}$, where $A_i = A$ for each $i \in I$. From Example 3.5, we have that B is an Esemihoop. For all $f \in B$, we define $\hat{s}(f) = s(f_1)$ on B. For any $f, g \in B$, there is $m \in \mathbf{I}(B)$ such that $f, g \leq m$. We can get that $\hat{s}(f) + \hat{s}(f \to_m g) = \hat{s}(g) + \hat{s}(g \to_m f)$. Thus, $\hat{s}$ is a Bosbach state on B.

Proposition 6.4. Let A be an Esemihoop with the least element 0 and s a Bosbach state. For all $x, y, z \in A$ and $a \in \mathbf{I}(A)$ such that $x \leq a$, the following statements hold.

(1) $s(x) \leq s(a)$ and $s(x^{-a}) = s(a) - s(x)$;

(2) $s(x^{-a-a}) = s(x)$;

(3) $s(x^{-a-a} \to_a x) = s(a)$;

Moreover, if $a \in \mathbf{I}(A)$ such that $x, y \leq a$,

(4) $x \leq y \implies s(x) \leq s(y)$;

(5) $s(x \to_a y) = s(y \to_a x) \iff s(x) = s(y)$;

(6) $s(x \to_a y^{-a-a}) = s(x^{-a-a} \to_a y) = s(x \to_a y)$;

(7) $s(x^{-a} \to_a y^{-a}) = s(y \to_a x)$;

(8) $s(x \odot y) = s(a) - s(x \to_a y^{-a})$;

(9) if $a \in \mathbf{I}(A)$ such that $x, y, z \leq a$, $s(x \to_a (y^{-a} \to_a z^{-a})) = s(x \to_a (z \to_a y))$.

Proof. (1) Since s is a Bosbach state on A, we have $s(x^{-a}) + s(x) = s(x \to_a 0) + s(x) = s(0) + s(0 \to_a x) = s(0) + s(a)$, which implies $s(x^{-a}) = s(a) - s(x)$. From $s(x^{-a}) \geq 0$, it follows $s(x) \leq s(a)$.

(2) By (1), $s(x^{-a-a}) = s(a) - s(x^{-a}) = s(a) - (s(a) - s(x)) = s(x)$.

(3) From (2) and Definition 6.1(3), we have $s(x) + s(x^{-a-a} \to_a x) = s(x^{-a-a}) + s(x^{-a-a} \to_a x) = s(x) + s(x \to_a x^{-a-a}) = s(x) + s(a)$. Thus, $s(x^{-a-a} \to_a x) = s(a)$.

(4) Suppose that $x \leq y$ and $a \in \mathbf{I}(A)$ such that $x, y \leq a$. We have $s(x) + s(a) = s(x) + s(x \to_a y) = s(y) + s(y \to_a x)$, that is, $s(x) - s(y) = s(y \to_a x) - s(a)$. From $y \to_a x \leq a$ and (1), we obtain $s(y \to_a x) \leq s(a)$. It proves that $s(x) - s(y) \leq 0$ and so $s(x) \leq s(y)$.

(5) Since $s(x) + s(x \to_a y) = s(y) + s(y \to_a x)$, this result is obvious.

(6) From $y \leq y^{-a-a}$, we have $x \to_a y \leq x \to_a y^{-a-a}$. Thus,

$$s((x \to_a y) \to_a (x \to_a y^{-a-a})) = s(a).$$

Since $y^{-a-a} \to_a y \leq (x \to_a y^{-a-a}) \to_a (x \to_a y)$, by (3) and (5), we get $s(a) = s(y^{-a-a} \to_a y) \leq s((x \to_a y^{-a-a}) \to_a (x \to_a y)) \leq s(a)$, which implies that $s((x \to_a y^{-a-a}) \to_a (x \to_a y)) = s(a)$. Hence,

$$s(x \to_a y^{-a-a}) + s(a) = s(x \to_a y^{-a-a}) + s((x \to_a y^{-a-a}) \to_a (x \to_a y))$$
$$= s(x \to_a y) + s((x \to_a y) \to_a (x \to_a y^{-a-a}))$$
$$= s(x \to_a y) + s(a).$$

This shows that $s(x \to_a y^{-a-a}) = s(x \to_a y)$. In a similar way, we can prove $s(x^{-a-a} \to_a y) = s(x \to_a y)$.

(7) Since $x^{-a} \to_a y^{-a} = y \to_a x^{-a-a}$ and (6), we have $s(x^{-a} \to_a y^{-a}) = s(y \to_a x^{-a-a}) = s(y \to_a x)$.

(8) It is easy to know that $s(x \odot y) + s((x \odot y)^{-a}) = s(0) + s(0 \to_a (x \odot y)) = s(a)$ and so $s(x \odot y) = s(a) - s((x \odot y)^{-a}) = s(a) - s(x \to_a y^{-a})$.

(9) Let $a \in \mathbf{I}(A)$ such that $x, y, z \leq a$. By (5), (7) and $z \to_a y \leq y^{-a} \to_a z^{-a}$, we have

$$s((y^{-a} \to_a z^{-a}) \to_a (z \to_a y)) = s((z \to_a y) \to_a (y^{-a} \to_a z^{-a})) = s(a).$$

Also, from Proposition 2.2(8), we have $(y^{-a} \to_a z^{-a}) \to_a (z \to_a y) \leq (x \to_a (y^{-a} \to_a z^{-a})) \to_a (x \to_a (z \to_a y)) \leq a$. Hence,

$$
\begin{aligned}
s(a) &= s((y^{-a} \to_a z^{-a}) \to_a (z \to_a y)) \\
&\leq s((x \to_a (y^{-a} \to_a z^{-a})) \to_a (x \to_a (z \to_a y))) \\
&\leq s(a).
\end{aligned}
$$

It proves that $s((x \to_a (y^{-a} \to_a z^{-a})) \to_a (x \to_a (z \to_a y))) = s(a)$. Similarly, we can get $s((x \to_a (z \to_a y)) \to_a (x \to_a (y^{-a} \to_a z^{-a}))) = s(a)$. Therefore, $s(x \to_a (y^{-a} \to_a z^{-a})) = s(x \to_a (z \to_a y))$ by (5). $\qquad \square$

Theorem 6.5. Let A be an Esemihoop with the least element 0. Suppose that $s : A \to [0, 1]$ satisfies $s(0) = 0$ and there exists $x_0 \in A$ such that $s(x_0) = 1$. For all $x, y \in A$ and $a \in \mathbf{I}(A)$ such that $x, y \leq a$, the following conditions are equivalent:
(1) s is a Bosbach state on A;
(2) $x \leq y \Longrightarrow s(y \to_a x) = s(a) + s(x) - s(y)$;
(3) $s(y \to_a x) = s(a) + s(x \wedge y) - s(y)$.

Proof. (1) $\Longrightarrow$ (2) Let $x \leq y$ and $a \in \mathbf{I}(A)$ such that $x, y \leq a$. Thus, we have $s(x) + s(a) = s(x) + s(x \to_a y) = s(y) + s(y \to_a x)$ and so $s(y \to_a x) = s(a) + s(x) - s(y)$.

(2) $\Longrightarrow$ (3) Since $x \wedge y \leq y$ and Proposition 2.2(10), we have $s(y \to_a x) = s(y \to_a (x \wedge y)) = s(a) + s(x \wedge y) - s(y)$.

(3) $\Longrightarrow$ (1) It follows from (3) that $s(y) + s(y \to_a x) = s(y) + s(a) + s(x \wedge y) - s(y) = s(a) + s(x \wedge y)$. For the same reason, $s(x) + s(x \to_a y) = s(a) + s(x \wedge y)$. Thus, we have $s(y) + s(y \to_a x) = s(x) + s(x \to_a y)$. It means that s is a Bosbach state. $\qquad \square$

Proposition 6.6. Let A be an Esemihoop with the least element 0 and s be a Bosbach state. Then
(1) $Ker(s) = \{x \in A | s(x) = 1\}$ is a proper filter of A;
(2) $(x, y) \in \theta_{Ker(s)} \Longleftrightarrow s(x) = s(y) = s(x \wedge y)$.

Proof. (1) Since s is a Bosbach state, there exists $x_0 \in A$ such that $s(x_0) = 1$. For all $x \in A$, there is $a \in \mathbf{I}(A)$ such that $x_0, x \leq a$. Thus, we have $1 = s(x_0) \leq s(a)$ and so $a \in Ker(s)$. It means that $Ker(s) \neq \emptyset$ and for all $x \in A$, there is $a \in \mathbf{I}(A) \cap Ker(s)$ such that $x \leq a$. Let $x, y \in A$ and $b \in \mathbf{I}(A)$ such that $x, y \leq b$. Suppose that $x, x \rightarrow_b y \in Ker(s)$. It follows from $x \leq y \rightarrow_b x$ that $1 = s(x) \leq s(y \rightarrow_b x)$. Hence, $s(y \rightarrow_b x) = 1$. From $s(x) + s(x \rightarrow_b y) = s(y) + s(y \rightarrow_b x)$, we have $s(y) = 1$, which implies $y \in Ker(s)$. From Proposition 4.11 and $s(0) = 0$, we have that $Ker(s) = \{x \in A | s(x) = 1\}$ is a proper filter of A.

(2) $\Longrightarrow$: Let $x, y \in A$ and $(x, y) \in \theta_{Ker(s)}$. There exists $a \in \mathbf{I}(A)$ such that $x, y \leq a$ and $x \rightarrow_a y, y \rightarrow_a x \in Ker(s)$. Thus, we have $s(x) = s(y)$ by Proposition 6.4(5). From $x \rightarrow_a y \leq a$, we get $1 = s(x \rightarrow_a y) \leq s(a)$ and so $s(a) = 1$. Therefore, $s(x \wedge y) = s(y \rightarrow_a x) - s(a) + s(y) = s(y)$.

$\Longleftarrow$: Suppose that $x, y \in A$ and $s(x) = s(y) = s(x \wedge y)$. For all $a \in \mathbf{I}(A)$ such that $x, y \leq a$, we have $s(y \rightarrow_a x) = s(a) + s(x \wedge y) - s(y) = s(a)$ and $s(x \rightarrow_a y) = s(a) + s(x \wedge y) - s(x) = s(a)$. Hence, $s(x \rightarrow_a y) = s(y \rightarrow_a x) = s(a)$. Since s is a Bosbach state, there is $x_0 \in A$ such that $s(x_0) = 1$. For each $b \in \mathbf{I}(A)$ such that $x_0, x, y \leq b$, we have $s(b) = 1$, which implies that $s(x \rightarrow_b y) = s(y \rightarrow_b x) = 1$ and so $x \rightarrow_b y, y \rightarrow_b x \in Ker(s)$. It proves that $(x, y) \in \theta_{Ker(s)}$. $\qquad \square$

Definition 6.7. Let A be an Esemihoop with the least element 0. Two elements $x, y \in A$ are said to be orthogonal if there is $a \in \mathbf{I}(A)$ such that $x, y \leq a$, $y^{-a-a} \leq x^{-a}$, and we write $x \perp y$.

Let A be an Esemihoop with the least element 0. If $x \perp y$, for all $a \in \mathbf{I}(A)$ such that $x, y \leq a$, we define $+_a$ on A: $x +_a y = x^{-a} \rightarrow_a y^{-a-a}$. It is clear that $x +_a y = x^{-a} \rightarrow_a y^{-a-a} = y^{-a} \rightarrow_a x^{-a-a} = y +_a x$.

Remark 6.8. Let A be an Esemihoop with the least element 0.

$$x \perp y \Longleftrightarrow \exists a \in \mathbf{I}(A) \text{ such that } x, y \leq a, \ y^{-a-a} \leq x^{-a}$$
$$\Longleftrightarrow \forall a \in \mathbf{I}(A) \text{ such that } x, y \leq a, \ y^{-a-a} \leq x^{-a}.$$

Let $a \in \mathbf{I}(A)$ such that $x, y \leq a$. By $y^{-a-a} \leq x^{-a}$, we have $x^{-a-a} \leq y^{-a-a-a} = y^{-a}$. For all $b \in \mathbf{I}(A)$ such that $x, y \leq b$, there is $c \in \mathbf{I}(A)$ such that $a, b \leq c$. By Proposition 3.9, we obtain $x \leq x^{-a-a} \leq y^{-a} \leq y^{-c}$. Thus, $x \leq y^{-c} \wedge b = y^{-b}$. Then in A_c, we get that $y^{-b} \rightarrow_c 0 \leq x \rightarrow_c 0$ and so $(y^{-b} \rightarrow_c 0) \wedge b \leq (x \rightarrow_c 0) \wedge b$. It proves that $y^{-b-b} \leq x^{-b}$. The other direction clearly holds.

Proposition 6.9. Let A be an Esemihoop with the least element 0. The following statements are equivalent: for all $x, y \in A$,
(1) $x \perp y$;

(2) for all $a \in \mathbf{I}(A)$ such that $x, y \leq a$, $x^{-a-a} \odot y^{-a-a} = 0$;
(3) $x \odot y = 0$.

Proof. $(1) \Longrightarrow (2) \Longrightarrow (3) \Longrightarrow (1)$ are easily proved. $\qquad \square$

Proposition 6.10. Let A be an Esemihoop with the least element 0. Then for all $x, y \in A$,
(1) $x \perp y \Longleftrightarrow y \perp x$;
(2) for any $a \in \mathbf{I}(A)$ such that $x \leq a$, $x \perp x^{-a}$ and $x \perp 0$;
(3) for all $a \in \mathbf{I}(A)$ such that $x, y \leq a$, $x \leq y \Longrightarrow x \perp y^{-a}$.

Proof. By Proposition 6.9, the proof is straightforward. $\qquad \square$

Definition 6.11. Let A be an Esemihoop with the least element 0. A map $s : A \to [0, 1]$ is called a Riečan state on A, if it satisfies the following conditions:
(1) there is $x_0 \in A$ such that $s(x_0) = 1$;
(2) if $x \perp y$, for all $a \in \mathbf{I}(A)$ such that $x, y \leq a$, $s(x +_a y) = s(x) + s(y)$.

We denote by $R[A]$ the set of all Riečan states on A.

Proposition 6.12. Let A be an Esemihoop with the least element 0 and s be a Riečan state. Then for all $x, y \in A$:
(1) $s(0) = 0$;
(2) for all $a \in \mathbf{I}(A)$ such that $x \leq a$, $s(x^{-a}) = s(a) - s(x)$ and $s(x^{-a-a}) = s(x)$;
(3) for all $a \in \mathbf{I}(A)$ such that $x \leq a$, $s(x) \leq s(a)$;
(4) $x \leq y \Longrightarrow s(x) \leq s(y)$;
(5) $Ker(s) = \{x \in A | s(x) = 1\}$ is a proper filter of A.

Proof. (1) Since s is a Riečan state, for all $a \in \mathbf{I}(A)$, we have $s(0 +_a 0) = s(0) + s(0)$. Also, we have $s(0 +_a 0) = s(0^{-a} \to_a 0^{-a-a}) = s(a \to_a 0) = s(0)$. It proves that $s(0) = 0$.

(2) Let $a \in \mathbf{I}(A)$ such that $x \leq a$. From $x^{-a} \perp x$ and $x^{-a} +_a x = x^{-a-a} \to_a x^{-a-a} = a$, we have $s(x^{-a} +_a x) = s(x^{-a}) + s(x)$ and $s(x^{-a} +_a x) = s(a)$. It follows that $s(x^{-a}) + s(x) = s(a)$ i.e. $s(x^{-a}) = s(a) - s(x)$. Moreover, $s(x^{-a-a}) = s(a) - s(x^{-a}) = s(a) - (s(a) - s(x)) = s(x)$.

(3) By (2), we have $s(x^{-a}) = s(a) - s(x)$. By $s(x^{-a}) \geq 0$, it follows $s(x) \leq s(a)$.

(4) Suppose that $x \leq y$ and $a \in \mathbf{I}(A)$ such that $x, y \leq a$. We have $x \perp y^{-a}$ by Proposition 6.10(3). Thus, $s(x +_a y^{-a}) = s(x) + s(y^{-a}) = s(x) + s(a) - s(y)$. It follows that $s(x) - s(y) = s(x +_a y^{-a}) - s(a)$. From (3), we have $s(x +_a y^{-a}) \leq s(a)$, which means that $s(x) \leq s(y)$.

(5) From the proof of Proposition 6.6(1), we have that $Ker(s) \neq \emptyset$ and (ESF1) holds. Let $x, y \in A$ and $b \in \mathbf{I}(A)$ such that $x, y \leq b$. Suppose that $x, x \to_b$

$y \in Ker(s)$. Clearly, $s(b) = 1$. It follows from $x \leq (x \to_b y^{-b-b}) \to_b y^{-b-b}$ and $x \to_b y \leq x \to_b y^{-b-b}$ that $s((x \to_b y^{-b-b}) \to_b y^{-b-b}) = s(x \to_b y^{-b-b}) = 1$. Hence, by (2), $s(x \odot y^{-b}) = s(b) - s((x \odot y^{-b})^{-b}) = 1 - s(x \to_b y^{-b-b}) = 0$. From $y^{-b-b} \leq x \to_b y^{-b-b} = (x \odot y^{-b})^{-b}$, we get $(x \odot y^{-b}) \perp y$. Thus, $s((x \odot y^{-b}) +_b y) = s(x \odot y^{-b}) + s(y) = s(y)$. We also have $s((x \odot y^{-b}) +_b y) = s((x \odot y^{-b})^{-b} \to_b y^{-b-b}) = s((x \to_b y^{-b-b}) \to_b y^{-b-b}) = 1$. Hence, we have $s(y) = 1$. It proves that $y \in Ker(s)$. Therefore, $Ker(s) = \{x \in A | s(x) = 1\}$ is a filter of A. By $s(0) = 0$, we get that $Ker(s)$ is proper. $\qquad\square$

Theorem 6.13. Let A be an Esemihoop with the least element 0. A Bosbach state on A is a Riečan state.

Proof. Let s be a Bosbach state on A and $x \perp y$. For all $a \in \mathbf{I}(A)$ such that $x, y \leq a$, we have $y^{-a-a} \leq x^{-a}$. It follows that $s(y^{-a-a} \to_a x^{-a}) = s(a)$. From $s(x^{-a}) + s(x^{-a} \to_a y^{-a-a}) = s(y^{-a-a}) + s(y^{-a-a} \to_a x^{-a})$, we have $(s(a) - s(x)) + s(x^{-a} \to_a y^{-a-a}) = s(y) + s(a)$ by Proposition 6.4. Then, $s(x +_a y) = s(x^{-a} \to_a y^{-a-a}) = s(x) + s(y)$. Therefore, s is a Riečan state. $\qquad\square$

Definition 6.14. Let A be an Esemihoop with the least element 0. A is called Glivenko if for all $a \in \mathbf{I}(A)$ such that $x, y \leq a$, $(x \to_a y)^{-a-a} = x \to_a y^{-a-a}$.

Obviously, a regular Esemihoop A is Glivenko.

Theorem 6.15. Let A be an Esemihoop with the least element 0. The following properties are equivalent: for all $x, y \in A$,
(1) A is Glivenko;
(2) for all $a \in \mathbf{I}(A)$ such that $x \leq a$, $(x^{-a-a} \to_a x)^{-a-a} = a$;
(3) for all $a \in \mathbf{I}(A)$ such that $x, y \leq a$, $(x \to_a y)^{-a-a} = x^{-a-a} \to_a y^{-a-a}$.

Proof. (1) $\implies$ (2) Let $a \in \mathbf{I}(A)$ such that $x \leq a$. Since A is Glivenko, $(x^{-a-a} \to_a x)^{-a-a} = x^{-a-a} \to_a x^{-a-a} = a$.

(2) $\implies$ (1) For all $a \in \mathbf{I}(A)$ such that $x, y \leq a$, $(y^{-a-a} \to_a y)^{-a-a} = a$. From $y^{-a-a} \to_a y \leq (x \to_a y^{-a-a}) \to_a (x \to_a y)$, we have

$$(y^{-a-a} \to_a y)^{-a-a} \leq ((x \to_a y^{-a-a}) \to_a (x \to_a y))^{-a-a}$$
$$\leq ((x \to_a y^{-a-a}) \to_a (x \to_a y)^{-a-a})^{-a-a}$$
$$= (x \to_a y^{-a-a}) \to_a (x \to_a y)^{-a-a} \text{ (Proposition 2.3)}.$$

This proves that $(x \to_a y^{-a-a}) \to_a (x \to_a y)^{-a-a} = a$ and so $x \to_a y^{-a-a} \leq (x \to_a y)^{-a-a}$. Since $x \to_a y \leq x \to_a y^{-a-a}$, we have $(x \to_a y)^{-a-a} \leq (x \to_a y^{-a-a})^{-a-a} = x \to_a y^{-a-a}$. Therefore, $(x \to_a y)^{-a-a} = x \to_a y^{-a-a}$. It follows that A is Glivenko.

(1) $\iff$ (3) For all $a \in \mathbf{I}(A)$ such that $x, y \leq a$, we get $x^{-a-a} \to_a y^{-a-a} = y^{-a} \to_a x^{-a} = (y^{-a} \odot x)^{-a} = x \to_a y^{-a-a}$. It is easy to check that A is Glivenko if and only if $(x \to_a y)^{-a-a} = x^{-a-a} \to_a y^{-a-a}$. $\qquad\square$

Theorem 6.16. Let A be a Glivenko Esemihoop. Bosbach states and Riečan states coincide on A.

Proof. Let s be a Riečan state. For all $x, y \in A$ and $a \in \mathbf{I}(A)$ such that $x, y \leq a$, we have $(x \wedge y)^{-a-a} \leq x^{-a-a}$. It follows that $x^{-a} \perp (x \wedge y)$. Thus, $s(x^{-a} +_a (x \wedge y)) = s(x^{-a}) + s(x \wedge y) = s(a) - s(x) + s(x \wedge y)$. Moreover,

$$
\begin{aligned}
s(x^{-a} +_a (x \wedge y)) &= s(x^{-a-a} \to_a (x \wedge y)^{-a-a}) \\
&= s((x \to_a (x \wedge y))^{-a-a}) \text{ (Theorem 6.15)} \\
&= s(x \to_a (x \wedge y)) \text{ (Proposition 6.12)} \\
&= s(x \to_a y) \text{ (Proposition 2.2).}
\end{aligned}
$$

Hence, $s(x \to_a y) = s(a) - s(x) + s(x \wedge y)$. By Theorem 6.5, s is a Bosbach state. This together with Theorem 6.13, we get that Bosbach states and Riečan states coincide on A. $\qquad\square$

Corollary 6.17. Bosbach states and Riečan states coincide on a regular Esemihoop A.

7 Internal States

In this section, we shall introduce the concept of internal states on Esemihoops. We show that there is a one-to-one correspondence between the set of all τ-compatible Riečan states on an Esemihoop A and the set of all Riečan states on $\tau(A)$. Furthermore, it is proved that the subset $\mathcal{T}$ of the power set of all prime state filters on an Esemihoop is a topological space.

Definition 7.1. Let A be an Esemihoop with the least element 0. An internal state on A is a mapping $\tau : A \to A$ such that the following conditions hold: for all $x, y \in A$,
(S1) $\tau(0) = 0$;
(S2) $x \leq y \implies \tau(x) \leq \tau(y)$;
(S3) for any $a \in \mathbf{I}(A)$ such that $x, y \leq a$, $\tau(x \to_a y) = \tau(x) \to_{\tau(a)} \tau(x \wedge y)$;
(S4) for any $a \in \mathbf{I}(A)$ such that $x, y \leq a$, $\tau(x \odot y) = \tau(x) \odot \tau(x \to_a (x \odot y))$;
(S5) $\tau(\tau(x) \odot \tau(y)) = \tau(x) \odot \tau(y)$;
(S6) $\tau(\tau(x) \wedge \tau(y)) = \tau(x) \wedge \tau(y)$.

Let A be an Esemihoop with the least element 0 and τ be an internal state on A. The pair (A, τ) is said to be a state Esemihoop. Two elements $x, y \in A$ are called comparable if $x \leq y$ or $y \leq x$.

Example 7.2. Let A be an Esemihoop with the least element 0. From Proposition 2.2(11), we can check that the identity $1_A : A \to A$ is an internal state on A.

Example 7.3. Let A and B be two Esemihoops with the least element 0. From Example 3.7, we have that $A \times B$ is an Esemihoop. Define a function τ on $A \times B$ as follows: $\tau((x_1, x_2)) = (x_1, 0)$ for all $(x_1, x_2) \in A \times B$. By Proposition 2.2 (10) and (11), we have that $\tau : A \times B \to A \times B$ is an internal state on $A \times B$.

Proposition 7.4. Let A be an Esemihoop with the least element 0 and τ be an internal state. The following properties hold: for all $x, y \in A$,
(1) if $a \in \mathbf{I}(A)$, $\tau(a) \in \mathbf{I}(A)$;
(2) for all $a \in \mathbf{I}(A)$ such that $x \leq a$, $\tau(x^{-a}) = (\tau(x))^{-\tau(a)}$;
(3) $\tau(x \odot y) \geq \tau(x) \odot \tau(y)$;
(4) for all $a \in \mathbf{I}(A)$ such that $x, y \leq a$, $\tau(x \to_a y) \leq \tau(x) \to_{\tau(a)} \tau(y)$. If x and y are comparable, $\tau(x \to_a y) = \tau(x) \to_{\tau(a)} \tau(y)$;
(5) $\tau^2(x) = \tau(x)$;
(6) $x \perp y \implies \tau(x) \perp \tau(y)$;
(7) for all $a \in \mathbf{I}(A)$ such that $x, y \leq a$, $x \perp y \implies \tau(\tau(x) +_{\tau(a)} \tau(y)) = \tau(x) +_{\tau(a)} \tau(y)$;
(8) $\tau(A) = \{x \in A | \tau(x) = x\}$.

Proof. (1) Let $a \in \mathbf{I}(A)$. From (S4), $\tau(a) = \tau(a \odot a) = \tau(a) \odot \tau(a \to_a (a \odot a)) = \tau(a) \odot \tau(a)$. It means that $\tau(a) \in \mathbf{I}(A)$.

(2) For all $a \in \mathbf{I}(A)$ such that $x \leq a$, by (S1) and (S3), $\tau(x^{-a}) = \tau(x \to_a 0) = \tau(x) \to_{\tau(a)} \tau(x \wedge 0) = \tau(x) \to_{\tau(a)} \tau(0) = \tau(x) \to_{\tau(a)} 0 = (\tau(x))^{-\tau(a)}$.

(3) For all $a \in \mathbf{I}(A)$ such that $x, y \leq a$, we have $y \leq x \to_a (x \odot y)$. It follows from (S2) that $\tau(y) \leq \tau(x \to_a (x \odot y))$. By (S4), $\tau(x) \odot \tau(y) \leq \tau(x) \odot \tau(x \to_a (x \odot y)) = \tau(x \odot y)$.

(4) Suppose that $a \in \mathbf{I}(A)$ such that $x, y \leq a$. Since $\tau(x \wedge y) \leq \tau(y)$, we get $\tau(x \to_a y) = \tau(x) \to_{\tau(a)} \tau(x \wedge y) \leq \tau(x) \to_{\tau(a)} \tau(y)$. If $x \leq y$, we obtain that $\tau(x) \leq \tau(y)$ and so $\tau(x) \to_{\tau(a)} \tau(y) = \tau(a)$. Also, we have $\tau(x \to_a y) = \tau(x) \to_{\tau(a)} \tau(x \wedge y) = \tau(x) \to_{\tau(a)} \tau(x) = \tau(a)$. Therefore, $\tau(x \to_a y) = \tau(x) \to_{\tau(a)} \tau(y)$. If $y \leq x$, then $\tau(x \to_a y) = \tau(x) \to_{\tau(a)} \tau(x \wedge y) = \tau(x) \to_{\tau(a)} \tau(y)$.

(5) Let $a \in \mathbf{I}(A)$ such that $x \leq a$. From (S6), $\tau^2(x) = \tau(\tau(x)) = \tau(\tau(x) \wedge \tau(a)) = \tau(x) \wedge \tau(a) = \tau(x)$.

(6) Assuming $x \perp y$. For all $a \in \mathbf{I}(A)$ such that $x, y \leq a$, we get $y^{-a-a} \leq x^{-a}$. This proves that $\tau(x), \tau(y) \leq \tau(a)$ and $\tau(y^{-a-a}) \leq \tau(x^{-a})$.
We have $(\tau(y))^{-\tau(a)-\tau(a)} \leq (\tau(x))^{-\tau(a)}$ by (2). Therefore, $\tau(x) \perp \tau(y)$.

(7) Let $x \perp y$. For all $a \in \mathbf{I}(A)$ such that $x, y \leq a$, we obtain that $\tau(y^{-a-a}) \leq \tau(x^{-a})$ and $\tau(x) \perp \tau(y)$ by (6). Thus, from (2), (4) and (5), we have

$$
\begin{aligned}
\tau[\tau(x) +_a \tau(y)] &= \tau((\tau(x))^{-\tau(a)} \to_{\tau(a)} (\tau(y))^{-\tau(a)-\tau(a)}) \\
&= \tau(\tau(x^{-a}) \to_{\tau(a)} \tau(y^{-a-a})) \\
&= \tau^2(x^{-a}) \to_{\tau(a)} \tau^2(y^{-a-a}) \\
&= \tau(x^{-a}) \to_{\tau(a)} \tau(y^{-a-a}) \\
&= (\tau(x))^{-\tau(a)} \to_{\tau(a)} (\tau(y))^{-\tau(a)-\tau(a)} \\
&= \tau(x) +_{\tau(a)} \tau(y).
\end{aligned}
$$

(8) For all $y \in \{x \in A | \tau(x) = x\}$, we have $y = \tau(y) \in \tau(A)$. Hence, $\{x \in A | \tau(x) = x\} \subseteq \tau(A)$. Conversely, let $y \in \tau(A)$. There is $x \in A$ such that $\tau(x) = y$. We have $\tau(y) = \tau(\tau(x)) = \tau(x) = y$. It follows that $y \in \{x \in A | \tau(x) = x\}$. Therefore, $\tau(A) = \{x \in A | \tau(x) = x\}$. $\qquad \square$

Proposition 7.5. Let (A, τ) be a state Esemihoop. $\tau(A)$ is a subalgebra of A.

Proof. Suppose that $\tau(x), \tau(y) \in \tau(A)$. From (S5) and (S6), $\tau(A)$ is closed under the operations $\wedge$ and $\odot$. There is $a \in \mathbf{I}(A)$ such that $x, y \leq a$. It follows $\tau(x), \tau(y) \leq \tau(a) \in \mathbf{I}(\tau(A))$. For all $a \in \mathbf{I}(A)$, we have $\tau(a) \in \mathbf{I}(A) \cap \tau(A)$. By Lemma 4.9 in [3], $\tau(A_a) = \{\tau(x) \in \tau(A) | x \leq a\}$ is a subalgebra of A_a. Therefore, $\tau(A)$ is a subalgebra of A. $\qquad \square$

Definition 7.6. Let (A, τ) be a state Esemihoop and s be a Riečan state on A. s is said to be τ-compatible if $\tau(x) = \tau(y)$ implies that $s(x) = s(y)$ for all $x, y \in A$.

The set of all τ-compatible Riečan state on (A, τ) is denoted by $R_\tau[A]$.

Theorem 7.7. Let (A, τ) be a state Esemihoop. There is a one-to-one correspondence between the set of all τ-compatible Riečan states on A and the set of all Riečan states on $\tau(A)$.

Proof. Assume that s is a Riečan state on $\tau(A)$. For all $x, y \in A$, we define a function $\varphi : R[\tau(A)] \to R_\tau[A]$ as follows: $\varphi(s)(x) = s(\tau(x))$. Next, we shall prove that the mapping φ is well defined. Since s is a Riečan state, there is $\tau(x_0) \in \tau(A)$ such that $s(\tau(x_0)) = 1$. Thus, $\varphi(s)(x_0) = s(\tau(x_0)) = 1$. Let $x \perp y$. For all $a \in \mathbf{I}(A)$ such that $x, y \leq a$, we get $y^{-a-a} \leq x^{-a}$. By Proposition 7.4(6), $\tau(x) \perp \tau(y)$. From

Proposition 7.4(4), we have

$$\begin{aligned}
\tau(x +_a y) &= \tau(x^{-a} \to_a y^{-a-a}) \\
&= \tau(x^{-a}) \to_{\tau(a)} \tau(y^{-a-a}) \\
&= (\tau(x))^{-\tau(a)} \to_{\tau(a)} (\tau(y))^{-\tau(a)-\tau(a)} \\
&= \tau(x) +_{\tau(a)} \tau(y).
\end{aligned}$$

It follows from $\tau(x) \perp \tau(y)$ that $\varphi(s)(x +_a y) = s(\tau(x +_a y)) = s(\tau(x) +_{\tau(a)} \tau(y)) = s(\tau(x)) + s(\tau(y)) = \varphi(s)(x) + \varphi(s)(y)$. Hence, $\varphi(s)$ is a Riečan state on A. If $\tau(x) = \tau(y)$, $\varphi(s)(x) = s(\tau(x)) = s(\tau(y)) = \varphi(s)(y)$. This proves that $\varphi(s)$ is a τ-compatible Riečan state on A.

Let s be a τ-compatible Riečan state on A. Define a mapping $\psi : R_\tau[A] \to R[\tau(A)]$ as follows $\psi(s)(\tau(x)) = s(x)$ for all $x, y \in A$. Now, we prove that $\psi(s)$ is a Riečan state on $\tau(A)$. Since s is a Riečan state on A, there is $x_0 \in A$ such that $s(x_0) = 1$. Then, $\psi(s)(\tau(x_0)) = s(x_0) = 1$. Suppose that $\tau(x) = \tau(y)$. We get $s(x) = s(y)$. If $\tau(x) \perp \tau(y)$, for all $\tau(a) \in \mathbf{I}(\tau(A))$ such that $\tau(x), \tau(y) \leq \tau(a)$, we have $(\tau(y))^{-\tau(a)-\tau(a)} \leq (\tau(x))^{-\tau(a)}$. It follows that $\tau(y^{-a-a}) \leq \tau(x^{-a})$. Therefore,

$$\begin{aligned}
\tau(\tau(x) +_{\tau(a)} \tau(y)) &= \tau((\tau(x))^{-\tau(a)} \to_{\tau(a)} (\tau(y))^{-\tau(a)-\tau(a)}) \\
&= \tau(\tau(x^{-a}) \to_{\tau(a)} \tau(y^{-a-a})) \\
&= \tau(x^{-a}) \to_{\tau(a)} \tau(y^{-a-a}) \\
&= (\tau(x))^{-\tau(a)} \to_{\tau(a)} (\tau(y))^{-\tau(a)-\tau(a)} \\
&= \tau(x) +_{\tau(a)} \tau(y).
\end{aligned}$$

Thus, we have

$$\begin{aligned}
\psi(s)(\tau(x) +_{\tau(a)} \tau(y)) &= \psi(s)(\tau(\tau(x) +_{\tau(a)} \tau(y))) \\
&= s(\tau(x) +_{\tau(a)} \tau(y)) \\
&= s(\tau(x)) + s(\tau(y)) \\
&= \psi(s)(\tau(\tau(x))) + \psi(s)(\tau(\tau(y))) \\
&= \psi(s)(\tau(x)) + \psi(s)(\tau(y)).
\end{aligned}$$

It means that $\psi(s)$ is a Riečan state on $\tau(A)$. The mapping ψ is well defined.

Suppose that s_1, s_2 are τ-compatible Riečan states on A and $\psi(s_1) = \psi(s_2)$. For all $x \in A$, $s_1(x) = \psi(s_1)(\tau(x)) = \psi(s_2)(\tau(x)) = s_2(x)$, which implies that $s_1 = s_2$. If s is a Riečan state on $\tau(A)$, for all $\tau(x) \in \tau(A)$, we obtain $s(\tau(x)) = \varphi(s)(x) = \psi(\varphi(s))(\tau(x))$. This proves that ψ is a bijection from $R_\tau[A]$ onto $R[\tau(A)]$ and $\psi^{-1} = \varphi$. Hence, there is a one-to-one correspondence between the set of τ-compatible Riečan states on A and the set of Riečan states on $\tau(A)$. $\qquad\square$

MIN LIU AND HONGXING LIU

Definition 7.8. Let (A, τ) be a state Esemihoop. A nonempty subset $F \subseteq A$ is said to be a state filter of (A, τ) if F is a filter of A and $x \in F$ implies that $\tau(x) \in F$ for all $x \in A$.

For all $\emptyset \neq F \subseteq A$, the smallest state filter of (A, τ) containing F is denoted by $\lfloor F \rfloor_\tau$, which is said to be the state filter of (A, τ) generated by F.

Theorem 7.9. Let (A, τ) be a state Esemihoop and $\emptyset \neq X \subseteq A$.
(1) $\lfloor X \rfloor_\tau = \{z \in A | z \geq (x_1 \odot \tau(x_1)) \odot \cdots \odot (x_n \odot \tau(x_n)), x_1, \cdots, x_n \in X, n \geq 1\}$;
(2) If $X = \{x\}$, $\lfloor x \rfloor_\tau = \{z \in A | z \geq (x \odot \tau(x))^n, n \geq 1\}$;
(3) If F is a state filter of (A, τ) and $x \notin F$, $\lfloor F \cup \{x\} \rfloor_\tau = \{z \in A | z \geq f \odot (x \odot \tau(x))^n, f \in F, n \geq 1\}$.

Proof. (1) Suppose that $S = \{z \in A | z \geq (x_1 \odot \tau(x_1)) \odot \cdots \odot (x_n \odot \tau(x_n)), x_1, \cdots, x_n \in X, n \geq 1\}$. Let $x \in X$. We have $x \geq x \odot \tau(x)$. It follows that $x \in S$ and so $X \subseteq S$. For all $x \in A$ and $x_1, \cdots, x_n \in X$, there is $a \in \mathbf{I}(A)$ such that $x, x_1, \cdots, x_n, \tau(x), \tau(x_1), \cdots, \tau(x_n) \leq a$. Thus, $a \geq (x_1 \odot \tau(x_1)) \odot \cdots \odot (x_n \odot \tau(x_n))$, which implies that $a \in S$. If $x, x \to_b y \in S$, where $b \in \mathbf{I}(A)$ with $x, y \leq b$, there are $m, n \geq 1$ and $x_1, \cdots, x_m, y_1, \cdots, y_n \in X$ such that $x \geq (x_1 \odot \tau(x_1)) \odot \cdots \odot (x_m \odot \tau(x_m))$ and $x \to_b y \geq (y_1 \odot \tau(y_1)) \odot \cdots \odot (y_n \odot \tau(y_n))$. Then $(x_1 \odot \tau(x_1)) \odot \cdots \odot (x_m \odot \tau(x_m)) \odot (y_1 \odot \tau(y_1)) \odot \cdots \odot (y_n \odot \tau(y_n)) \leq x \odot (x \to_b y) \leq y$. It means that $y \in S$. By Proposition 4.11, S is a filter of A containing X. For all $x \in S$, there exist $m \geq 1$ and $x_1, \cdots, x_m \in X$ such that $x \geq (x_1 \odot \tau(x_1)) \odot \cdots \odot (x_m \odot \tau(x_m))$. We get

$$\tau(x) \geq \tau((x_1 \odot \tau(x_1)) \odot \cdots \odot (x_m \odot \tau(x_m)))$$
$$\geq \tau((x_1 \odot \tau(x_1))) \odot \cdots \odot \tau((x_m \odot \tau(x_m)))$$
$$\geq (\tau(x_1) \odot x_1) \odot \cdots \odot (\tau(x_m \odot x_m).$$

So $\tau(x) \in S$. This proves that S is a state filter of (A, τ) containing X. Assume that T is a state filter of (A, τ) containing X. For all $x \in S$, there exist $m \geq 1$ and $x_1, \cdots, x_m \in X$ such that $x \geq (x_1 \odot \tau(x_1)) \odot \cdots \odot (x_m \odot \tau(x_m))$. From $x_1, \cdots, x_m \in X \subseteq T$, we get $\tau(x_1), \cdots, \tau(x_m) \in T$. We have $(x_1 \odot \tau(x_1)) \odot \cdots \odot (x_m \odot \tau(x_m)) \in T$, which means that $x \in T$ and so $S \subseteq T$. Therefore $\lfloor X \rfloor_\tau = \{z \in A | z \geq (x_1 \odot \tau(x_1)) \odot \cdots \odot (x_n \odot \tau(x_n)), x_1, \cdots, x_n \in X, n \geq 1\}$.
The proofs of (2) and (3) are similar to (1). $\qquad \square$

Proposition 7.10. Let (A, τ) be a state Esemihoop and $x, y \in A$.
(1) $x \leq y \Longrightarrow \lfloor y \rfloor_\tau \subseteq \lfloor x \rfloor_\tau$;
(2) $\lfloor \tau(x) \rfloor_\tau \subseteq \lfloor x \rfloor_\tau = \lfloor x \odot \tau(x) \rfloor_\tau$.

Proof. The proof is straightforward. □

Definition 7.11. Let (A, τ) be a state Esemihoop. A proper state filter F of (A, τ) is said to be prime if for any two state filter F_1, F_2, $F_1 \cap F_2 \subseteq F \Longrightarrow F_1 \subseteq F$ or $F_2 \subseteq F$.

We denote the set of all prime state filters of (A, τ) by $\mathrm{PSF}[A]$.

Definition 7.12. Let (A, τ) be a state Esemihoop and F a proper state filter. F is said to be maximal if no proper state filter of (A, τ) can strictly contain it.

Lemma 7.13. Let (A, τ) be a state Esemihoop and F a proper state filter. The following conditions are equivalent:
(1) F is a maximal state filter;
(2) for all $x \in A \backslash F$, $\lfloor F \cup \{x\} \rfloor_\tau = A$;
(3) for all $x \in A \backslash F$, there are $a \in \mathbf{I}(A)$ and a positive integer $n \geq 1$ such that $x \leq a$, $((\tau(x))^n)^{-\tau(a)} \in F$.

Proof. (1) $\Longleftrightarrow$ (2) The proof is similar to Theorem 5.3.
    (2) $\Longrightarrow$ (3) Suppose that $x \in A \backslash F$. We have $0 \in A = \lfloor F \cup \{x\} \rfloor_\tau$. There are $f \in F$ and $n \in \mathbb{N}^*$ such that $0 \geq f \odot (x \odot (\tau(x)))^n$. It follows that $0 = \tau(0) \geq \tau(f) \odot \tau((x \odot (\tau(x)))^n) \geq \tau(f) \odot (\tau(x))^{2n}$. There exists $a \in \mathbf{I}(A)$ such that $x, f \leq a$ and $\tau(f) \leq ((\tau(x))^{2n})^{-\tau(a)}$. From $\tau(f) \in F$, we get $((\tau(x))^{2n})^{-\tau(a)} \in F$.
    (3) $\Longrightarrow$ (2) Let $x \in A \backslash F$. There are $a \in \mathbf{I}(A)$ and a positive integer $n \geq 1$ such that $x \leq a$ and $((\tau(x))^n)^{-\tau(a)} \in F$. Then $((\tau(x))^n)^{-\tau(a)} \odot (x \odot \tau(x))^n \leq ((\tau(x))^n)^{-\tau(a)} \odot (\tau(x))^n = 0$. This together with $((\tau(x))^n)^{-\tau(a)} \in F$ implies $0 \in \lfloor F \cup \{x\} \rfloor_\tau$. Thus, $\lfloor F \cup \{x\} \rfloor_\tau = A$. □

Let (A, τ) be a state Esemihoop and $X \subseteq A$. We define $[X] = \{F \in \mathrm{PSF}[A] | X \not\subseteq F\}$, which is a subset of $\mathrm{PSF}[A]$. If $X = \{x\}$, let $[x] = [\{x\}] = \{F \in \mathrm{PSF}[A] | x \notin F\}$.

Proposition 7.14. Let (A, τ) be a state Esemihoop, $X, Y \subseteq A$ and $\{X_i\}_{i \in I}$ be family subsets of A. The following properties hold:
(1) $X \subseteq Y \Longrightarrow [X] \subseteq [Y]$;
(2) $[0] = \mathrm{PSF}[A]$, $[\emptyset] = \emptyset$;
(3) $[X] \cap [Y] = [\lfloor X \rfloor_\tau \cap \lfloor Y \rfloor_\tau]$;
(4) $\bigcup_{i \in I}[X_i] = [\bigcup_{i \in I} X_i]$;
(5) $[X] = [\lfloor X \rfloor_\tau]$.

Proof. The proofs of (1) and (2) are straightforward.
    (3) Let $F \in [X] \cap [Y]$. We get $X, Y \not\subseteq F$. It follows that $\lfloor X \rfloor_\tau, \lfloor Y \rfloor_\tau \not\subseteq F$ and $\lfloor X \rfloor_\tau \cap \lfloor Y \rfloor_\tau \not\subseteq F$. This means that $F \in [\lfloor X \rfloor_\tau \cap \lfloor Y \rfloor_\tau]$. Hence, $[X] \cap [Y] \subseteq$

$\lfloor \lfloor X \rfloor_\tau \cap \lfloor Y \rfloor_\tau \rfloor$. Conversely, if $F \in [\lfloor X \rfloor_\tau \cap \lfloor Y \rfloor_\tau]$, we have $\lfloor X \rfloor_\tau \cap \lfloor Y \rfloor_\tau \not\subseteq F$. We have that $X \subseteq \lfloor X \rfloor_\tau \not\subseteq F$ and $Y \subseteq \lfloor Y \rfloor_\tau \not\subseteq F$. Thus, $F \in [X] \cap [Y]$. This proves $[\lfloor X \rfloor_\tau \cap \lfloor Y \rfloor_\tau] \subseteq [X] \cap [Y]$.

(4) Since $X_i \subseteq \bigcup_{i \in I} X_i$ for all $i \in I$, we have $[X_i] \subseteq [\bigcup_{i \in I} X_i]$. Obviously, $\bigcup_{i \in I} [X_i] \subseteq [\bigcup_{i \in I} X_i]$. Assume that $F \in [\bigcup_{i \in I} X_i]$. There is $j \in I$ such that $X_j \not\subseteq F$, which means that $F \in [X_j]$ and $F \in \bigcup_{i \in I} [X_i]$. Thus, $[\bigcup_{i \in I} X_i] \subseteq \bigcup_{i \in I} [X_i]$.

(5) Since $F \in [X] \iff X \not\subseteq F \iff \lfloor X \rfloor_\tau \not\subseteq F \iff F \in [\lfloor X \rfloor_\tau]$, we get $[X] = [\lfloor X \rfloor_\tau]$. $\qquad\square$

Proposition 7.15. Let (A, τ) be a state Esemihoop and $x, y \in A$. Then
(1) $[x] = [\lfloor x \rfloor_\tau]$;
(2) $x \le y \implies [y] \subseteq [x]$;
(3) $[x] \cup [y] = [x \odot y]$.

Proof. The proof is straightforward. $\qquad\square$

Theorem 7.16. Let (A, τ) be a state Esemihoop and $\mathcal{T} = \{[X] | X \subseteq A\}$ a subset of the power set of PSF$[A]$. Then $\mathcal{T}$ is a topology on PSF$[A]$.

Proof. By Proposition 7.14(2), (3) and (4), the proof is obvious. $\qquad\square$

8 Conclusion

In this paper, we give the notion of Esemihoops, which is an extension of semihoops and Ehoops. We also present some properties of Esemihoops. Moreover, we define congruences, ideals and filters on Esemihoops. The relations between them are discussed. Because Esemihoops do not satisfy $x \odot (x \to y) = y \odot (y \to x)$ and $x \wedge a = x \odot a$ for all $a \in \mathbf{I}(A)$, the proofs of many properties are different from those of Ehoops. We have that every Esemihoop has at least one maximal filter. In a regular Esemihoop A, we prove that A has at least one maximal ideal and every proper ideal is contained in a maximal ideal. In addition, it is proved that a Bosbach state on Esemihoops is a Riečan state. Specially, Bosbach states and Riečan states are consistent on Glivenko Esemihoops. Furthermore, we show that there is a one-to-one correspondence between the set of τ-compatible Riečan states on A and the set of Riečan states on $\tau(A)$. Moreover, we get that the subset $\mathcal{T}$ of the power set of all prime state filters on Esemihoop is a topological space.

References

[1] F. Esteva, L. Godo, P. Hájek, F. Montagna. (2003). *Hoops and fuzzy logic*, Journal of Logic and Computation, 13, 532–555.

[2] R. A. Borzooei, M. Aaly Kologani. (2015). *Local and perfect semihoops*, Journal of Intelligent and Fuzzy Systems, 29, 223–234.

[3] P. He, B. Zhao, X. Xin. (2016). *States and internal states on semihoops*, Soft Computing, 21, 2941–2957.

[4] P. He, J. Wang, J. Yang. (2022). *The existence of states based on Glivenko semihoops*, Mathematical Logic, 61, 1145–1170.

[5] B. Bosbach. (1969). *Komplementäre Halbgruppen. Axiomatik und Arithmetik*, Fundamenta Mathematicae, 64, 257–287.

[6] A. Dvurečenskij, O. Zahiri. (2019). *On EMV-algebras*, Fuzzy sets and systems, 373, 116–148.

[7] H. Liu. (2020). *EBL-algebras*, Soft Computing, 24, 14333–14343.

[8] H. Niu, X. Xin, J. Wang. (2020). *Ideal theory on bounded semihoops*. Italian Journal of Pure and Applied Mathematics, 44, 911–925.

[9] F. Xie, H. Liu. (2021). *Ehoops*. Journal of Multiple–Valued Logic and Soft Computing, 37, 77–106.

[10] Y. Fu, X. Xin, J. Wang. (2018). *State maps on semihoops*. Open Mathematics, 16(1), 1061–1076.

[11] L. Zhang, X. Xin. (2019). *Derivations and differential filters on semihoops*. Italian Journal of Pure and Applied Mathematics, 42, 916–933.

[12] Y. Tang, X. Xin, X. Zhou. (2022). *Spectra and reticulation of semihoops*. Open Mathematics, 20, 1276–1287.

[13] J. Wang, T. Qian, Y. She. (2019). *Characterizations of obstinate filters in semihoops*. Italian Journal of Pure and Applied Mathematics, 42, 851–862.

[14] J. Wang, G. Wang, M. Hu. (2008). *Topology on the set of R_0 semantics for R_0 algebras*. Soft Computing, 12, 585–591.

 Received 9 July 2023

Formalization of the Telegrapher's Equations using Higher-Order-Logic Theorem Proving

Elif Deniz[1], Adnan Rashid[1,2], Osman Hasan[2] and Sofiène Tahar[1]

[1]*Department of Electrical and Computer Engineering*
Concordia University, Montreal, QC, Canada
`{e_deniz, rashid, tahar}@ece.concordia.ca`

[2]*School of Electrical Engineering and Computer Science*
National University of Sciences and Technology, Islamabad, Pakistan
`osman.hasan@seecs.nust.edu.pk`

Abstract

The telegrapher's equations constitute a set of linear partial differential equations that establish a mathematical correspondence between the electrical current and voltage within transmission lines, taking into account factors, such as distance and time. These equations find wide applications in the design and analysis of various systems, including integrated circuits and antennas. This paper proposes the utilization of higher-order-logic theorem proving for a formal analysis of the telegrapher's equations, also referred to as the transmission line equations. Specifically, we present a formal model of the telegrapher's equations in both time and phasor domains. Subsequently, we employ the HOL Light theorem prover to formally verify the solutions of the telegrapher's equations in the phasor domain. Furthermore, we established a connection between phasor and time-domain functions to formally verify the general solutions for the time-domain partial differential equations for the current and voltage in an electric transmission line. To demonstrate the practical effectiveness of our proposed formalization, we conduct a formal analysis of a terminated transmission line and its special cases, i.e., short- and open-circuited transmission lines commonly used in antenna design, by formally verifying the load impedance and the voltage reflection coefficient.

1 Introduction

Transmission line theory provides a fundamental framework for understanding and analyzing the behavior of transmission lines in the context of their application in various domains, such as integrated circuits and antennas. It serves as a mathematical foundation, capturing an efficient power transfer, ensuring dependable communication, optimizing system design and achieving an electromagnetic compatibility. Therefore, electrical transmission lines play a pivotal role in the conveyance of signals and electrical energy, primarily for the transmission of power, from a source to a load. For instance, in real life, a transmission line acts as a conduit for distributing electricity to homes, businesses, industries, and hospitals, working in conjunction with power generation plants and substations. Sometimes a disruption in the power supply resulting from a transmission line breakage can lead to serious consequences. For instance, in a hospital environment, power failures directly jeopardize the safety of patients and medical personnel. Consequently, emphasizing the crucial importance of guaranteeing the dependability of the electrical elements in the transmission line is essential.

Transmission lines are comprised of a minimum of two conductors that facilitate an efficient and a reliable transmission of information and energy. A two-conductor transmission line supports a *transverse electromagnetic* (TEM) wave [1], where the electric and magnetic fields are perpendicular to each other and transverse to the direction of propagation of waves along the transmission line. TEM waves have a fundamental property of establishing a distinct relationship between the electric $\mathbf{E}$ and the magnetic $\mathbf{H}$ fields, which are specifically related to the voltage V and current I, respectively as the following Maxwell's equations:

$$V = - \int_L \mathbf{E}.d\mathbf{l}, \tag{1}$$

$$I \; = \; \oint_L \mathbf{H}.d\mathbf{l} \tag{2}$$

The analysis of transmission lines can be made simpler by only focusing on the circuit quantities, V and I, rather than directly solving the complex line integral based Maxwell's equations (Equations (1) and (2)) and boundary conditions involving electric and magnetic fields ($\mathbf{E}$ and $\mathbf{H}$). In this regard, we employ an equivalent circuit in order to represent the transmission line's behavior. The purpose of developing an equivalent circuit model is to simplify the intricate electromagnetic interactions inherent to the transmission line, thereby reducing them to a set of lumped elements amenable to analysis through circuit theory.

Following the construction of the equivalent circuit, the telegrapher's, also referred to as the transmission line equations, can be derived using circuit analysis techniques. The behavior of transmission lines is elucidated through the utilization of the telegrapher's equations that are based on Partial Differential Equations (PDEs) and rigorously capture the complex electromagnetics and propagation dynamics occurring within these transmission systems. Next, by applying appropriate boundary conditions and simplification of assumptions, the telegrapher's equations provide a useful mathematical model for analyzing transmission lines. Furthermore, comprehending and analyzing their solutions is crucial to ensure the reliability and safety of our everyday electrical systems. For instance, solutions derived from the telegrapher's equations can be seamlessly integrated into the modeling of signal processing and communications systems, such as filters, matching networks, transmission lines, transformers, and small-signal models for transistors. Thus, our goal is to establish foundational concepts rooted in the telegrapher's equations, aiming to subsequently expand this basis for the analysis of more complex engineering applications.

There are numerous analytical and numerical approaches that have been used to solve the PDE based transmission line equations. For example, finite differences [2] and iterative [3] methods are a few numerical approaches that are applied to obtain the solution of these equations. These numerical techniques are highly efficient approaches as they use recursive algorithms to find out the solution of these PDEs. However, due to the finite precision of computer arithmetic and the involvement of round off approximations, these methods cannot guarantee the accuracy of the analysis. On the other hand, analytical solutions may provide a complementary point of view by deriving a closed-form exact solution. However, such analysis is usually done using paper-and-pencil proof methods and is hence prone to human error, especially, for larger systems. Therefore, these conventional methods cannot be trusted to provide accurate analysis, in particular for safety-critical applications.

Several approaches have been used to find the solutions of PDEs for the analysis of the telegrapher's equations. For instance, Konane et al. [4] proposed an exact solution of the telegrapher's equations for voltage monitoring of electrical transmission lines. Kühn [5] developed a general solution of the telegrapher's equations for electrically short transmission lines based on circuit theory. Similarly, Biazar et al. [3] proposed an iterative method to obtain an approximate solution of the telegrapher's equation. However, all these contributions are based on traditional analysis methods.

Formal methods, in particular interactive theorem proving, have also been used for analyzing other forms of PDEs. For example, Boldo et al. [6] formally verified the numerical solution of the wave equation [7] using the Coq theorem prover[1]. Similarly, Deniz et al. [8] formalized the one-dimensional heat equation [9] and verified the general solution of the equation and its convergence in the HOL Light theorem prover[2]. However, none of the aforementioned contributions focused on the telegrapher's equations. In this paper, we present a framework for formally analyzing the telegrapher's equations and their analytical solutions within higher-order-logic theorem proving. We first provide the formal definitions of the telegrapher's equations and their alternate representations, i.e., the wave equations both in the time and phasor domains by proving the relationship between these equations in the phasor domain. We also develop the reasoning steps for the verification of the analytical solutions of these equations, which, to the best of our knowledge, are not available in other theorem provers. In addition, we prove some important properties of special types of transmission line which are lossless and distortionless. In order to demonstrate the utilization of our work, we formally analyze the terminated, short- and open circuited transmission lines. We opted to use the HOL Light theorem prover for the proposed formalization of the telegrapher's equations due to the availability of rich libraries of the multivariate calculus. The HOL Light code developed in this paper is available at [10].

The rest of the paper is structured as follows: We present the proposed framework for the formalization of the telegrapher's equations in higher-order-logic in Section 2. Section 3 describes some preliminary details of the multivariate libraries of the HOL Light theorem prover that are necessary for understanding the rest of the paper. We present the formalization of the telegrapher's equations and a derived form of the wave equations in time and phasor domains alongside a verification of their relationship in the phasor domain in Section 4. In Section 5, we provide the formal verification of the analytical solutions of the telegrapher's equations. Section 6 provides the formal analysis of a terminated, short-circuited and open-circuited transmission lines that illustrate the practical effectiveness of our proposed formalizations. We discuss the difficulties encountered during our work and gained experience in Section 7. Finally, Section 8 concludes the paper.

2 Proposed Methodology

The proposed approach for formally analyzing the telegrapher's equations and their derived form (the wave equations) using higher-order-logic theorem proving is depicted in Figure 1.

---

[1]https://coq.inria.fr/
[2]https://www.cl.cam.ac.uk/jrh13/hol-light/

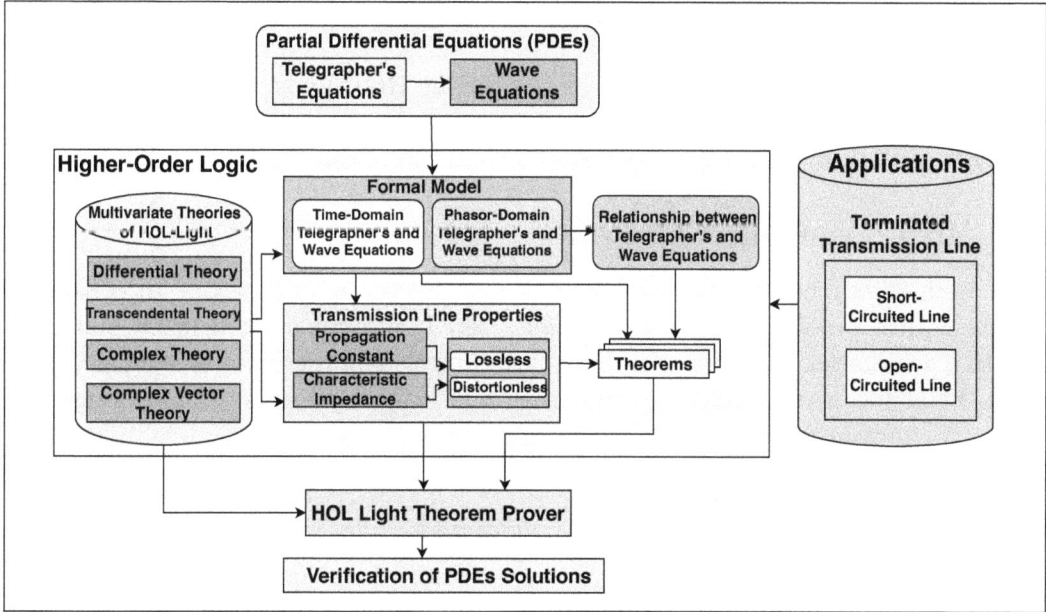

Figure 1: Proposed Methodology

The first step of our proposed approach is to formalize the telegrapher's and the wave equations in time and phasor domains. The formalization of these equations requires HOL Light's libraries of multivariate calculus, such as differential, transcendental and complex vectors. The next step is to establish theorems that enable the formal verification of solutions for these equations by leveraging the advantages of the phasor-domain representation of these equations, which simplifies the time-domain PDEs. Moreover, the relationship between the telegrapher's and the wave equations in the phasor domain is formally verified using these theorems. Subsequently, we use the solutions in the phasor domain to verify the PDEs by establishing a relationship between the corresponding functions in the phasor and time domains. All theorems of the proposed framework of the telegrapher's equations are verified in HOL Light in a generic way in order to obtain general (universally quantified) solutions of the related PDEs. The next step is to represent some important properties of transmission lines, such as the propagation constant and the characteristic impedance specifically focusing on the case of lossless and distortionless lines. Moreover, in order to demonstrate the practical effectiveness of the proposed formalization, we conduct a formal analysis of terminated transmission line and its special cases short-circuited and open-circuited tranmission lines, which are extensively used in electrical and telecommunication systems.

3 Preliminaries

In this section, we provide a brief overview of the HOL Light theorem prover, the HOL Light functions and symbols and some definitions from the theory of complex analysis of HOL Light that are necessary for understanding the rest of the paper.

3.1 HOL Light Theorem Prover

The HOL Light theorem prover [11] is a mechanized proof-assistant to construct mathematical proofs in higher-order-logic [12]. It is implemented in OCaml [13], which is a variant of the ML (Meta-Language) functional programming language [14]. HOL Light has a very small logical kernel, which includes some basic axioms and primitive inference rules.

HOL Light Symbols	Standard Symbols	Description
@x.t(x)	εx. t(x)	Some x such that t(x) is true
&a	$\mathbb{N} \to \mathbb{R}$	Type casting from Natural numbers to Reals
&num	$\{0, 1, 2..\}$	Positive Integers data type
Cx(a)	$\mathbb{R} \to \mathbb{C}$	Type casting from Reals to Complex
real	$\mathbb{R}$	Real data type
complex	$\mathbb{C}$	Complex data type
csqrt x	$\sqrt{x}$	Complex square root function

Table 1: HOL Light Symbols

Soundness is guaranteed by ensuring that every new theorem is verified by applying these basic axioms and inference rules or any other previously verified theorems/inference rules. In HOL Light, which is based on classical logic, a *theory* comprises types, constants, axioms, definitions, and theorems. HOL supports two interactive proof methods: forward and backward. In a forward proof, users begin with theorems that have already been proven and apply inference rules to arrive at the desired theorem. On the other hand, a backward or goal-directed proof method is the opposite of the forward approach. It relies on the concept of *tactics*, which are OCaml functions that reduce the goals into more manageable subgoals, which are verified to conclude with the proofs of theorems. Furthermore, HOL Light contains lemmas, which are proved as part of the more extensive proof process for theorems. The user can choose to either utilize established lemmas or prove new lemmas as they work towards their main objective of proving the theorems. One of the important features of HOL Light is the availability of many automatic proof procedures that help users in conducting proofs in an efficient manner. Table 1 summarizes some HOL functions and symbols and their meanings that are used in this paper.

3.2 Complex Analysis Library

We now present some of the common HOL Light functions that are used in the proposed analysis.

Definition 3.1. *Re and Im*
$\vdash_{def}$ ∀z. Re z - z$1
$\vdash_{def}$ ∀z. Im z = z$2

The functions `Re` and `Im` represent the real and imaginary parts of a complex number, respectively. Here, the notation `z$i` represents the ith component of a vector `z`.

Definition 3.2. *Cx and ii*
$\vdash_{def}$ ∀a. Cx a = complex (a, &0)
$\vdash_{def}$ ii = complex (&0, &1)

`Cx` is a type casting function with a data-type $\mathbb{R} \to \mathbb{C}$. It accepts a real number and returns its corresponding complex number with the imaginary part as zero. Also, the types $\mathbb{R}^2$ and $\mathbb{C}$ are synonymous. The `&` operator has data-type $\mathbb{N} \to \mathbb{R}$ and is used to map a natural number to a real number. Similarly, the function `ii` (iota) represents a complex number with a real part equal to 0 and the magnitude of the imaginary part equal to 1. In our formalization, the symbol `ii` is employed to represent j denoting the imaginary number.

Definition 3.3. *Exponential Functions*
$\vdash_{def}$ ∀x. exp x = Re (cexp (Cx x))

The HOL Light functions `exp` and `cexp` with data-types $\mathbb{R} \to \mathbb{R}$ and $\mathbb{C} \to \mathbb{C}$ represent the real and complex exponential functions, respectively.

Definition 3.4. *Complex Derivative*
$\vdash_{def}$ ∀f x. complex_derivative f x =
          (@f'. (f has_complex_derivative f') (at x))

The function `complex_derivative` describes the complex derivative in functional form. It accepts a function `f`: $\mathbb{C} \to \mathbb{C}$ and a complex number `x`, which is the point at which `f` has to be differentiated, and returns a variable of data-type $\mathbb{C}$, providing the derivative of `f` at `x`. Here, the term `at` indicates a specific point at which the differentiation is being evaluated, namely, at the value of `x`.

Definition 3.5. *Higher Complex Derivative*

$\vdash_{def} \forall$f x.
higher_complex_derivative 0 f x = f x $\wedge$
($\forall$n. higher_complex_derivative (SUC n) f x
= (complex_derivative (λx. higher_complex_derivative n f x) x))

The function `higher_complex_derivative` represents the nth-order derivative of the function f. It accepts an order n of the derivative, a function f: $\mathbb{C} \to \mathbb{C}$ and a complex number x, and provides the nth derivative of f at x.

To facilitate in the comprehension of the paper to a non-HOL user, we articulate the telegrapher's equations and the associated lemmas through a blend of Math/HOL Light notation, and some of the frequently used functions in our formalization, their meaning and the associated mathematical conventions are presented in Table 2.

HOL Light Functions	Mathematical Conventions	Description
cexp x	$\overrightarrow{e}^x$	Complex exponential function
ctan	$\overrightarrow{\tan}$	Tangent of a complex-valued function
complex_derivative (λz. V(z)) z	$\dfrac{\overrightarrow{dV(z)}}{dz}$	Derivative of a complex-valued function V w.r.t z
higher_complex_derivative 2 (λz. V(z)) z	$\dfrac{\overrightarrow{d^2V(z)}}{dz^2}$	Second-order derivative of a complex-valued function V w.r.t z
complex_derivative (λz. V z t) z	$\dfrac{\overrightarrow{\partial V(z,t)}}{\partial z}$	Partial derivative of a complex-valued function V w.r.t z
higher_complex_derivative 2 (λz. V z t) z	$\dfrac{\overrightarrow{\partial^2 V(z,t)}}{\partial z^2}$	Second-order partial derivative of a complex-valued function V w.r.t z

Table 2: Conventions used for HOL Light Functions

4 Formalization of the Telegrapher's Equations

The telegrapher's equations are a pair of coupled linear PDEs that describe how the voltage and current change along a transmission line with respect to distance and time. Figure 2 depicts an equivalent circuit model of a two-conductor transmission line. Here, R represents the line parameter resistance, whereas the other line parameters are the inductance L, the capacitance C, and the conductance G, which are specified per unit length (Δz). Moreover, $V(z,t)$ and $V(z+\Delta z,t)$ are the input and output voltages, respectively. Similarly, $I(z,t)$ and $I(z+\Delta z,t)$ are the input and output currents, respectively. Moreover, both voltage and current are functions of space and time.

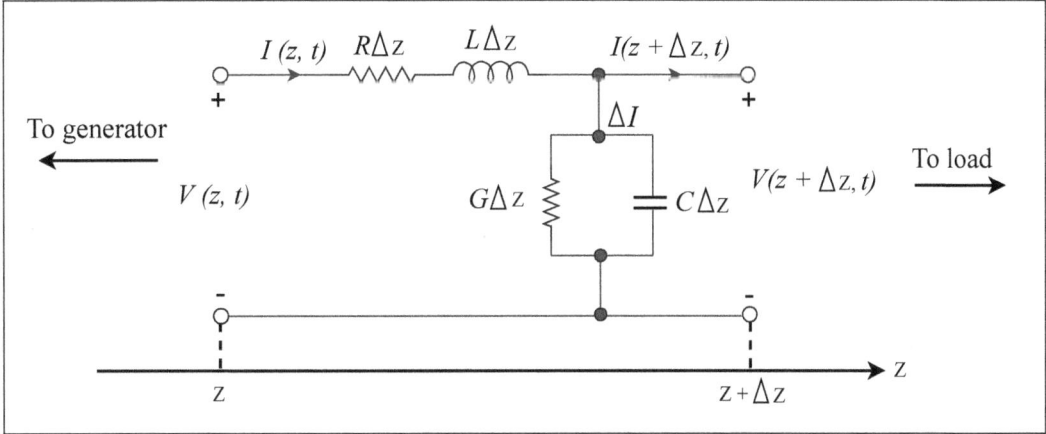

Figure 2: Equivalent Circuit of Two-Conductor Transmission Line [15]

4.1   Telegrapher's Equations in Time Domain

The Law of Conservation of Energy, attributed to Kirchhoff, asserts that there is no loss of voltage throughout a closed loop or circuit; instead, one returns to the initial point within the circuit and, consequently, to the same initial electric potential. Hence, any reductions in voltage within the circuit must balance out with the voltage sources encountered along the same route. By applying the Kirchhoff's voltage law to the circuit of two-conductor transmission line of Figure 2, we get the following equations [16]:

$$V(z+\Delta z,t) - V(z,t) = -R\Delta z I(z,t) - L\Delta z \frac{\partial I(z,t)}{\partial t} \qquad (3)$$

Next, dividing Equation (3) by Δz and applying the limit $\Delta z \to 0$, we obtain:

$$\lim_{\Delta z \to 0} \frac{V(z+\Delta z,t) - V(z,t)}{\Delta z} = \frac{-R\Delta z I(z,t)}{\Delta z} - L\frac{\Delta z}{\Delta z}\frac{\partial I(z,t)}{\partial t}$$

Finally, by using the definition of the partial derivative, we get:

$$\frac{\partial V(z,t)}{\partial z} = -RI(z,t) - L\frac{\partial I(z,t)}{\partial t} \qquad (4)$$

205

Similarly, by applying the Kirchhoff's current law to the circuit, we find [16]:

$$I(z + \Delta z, t) - I(z, t) = -G\Delta z V(z + \Delta z, t) - C\Delta z \frac{\partial V(z + \Delta z, t)}{\partial t} \tag{5}$$

Next, dividing Equation (5) by Δz and using the definition of the partial derivative, we get:

$$\frac{\partial I(z, t)}{\partial z} = -GV(z, t) - C\frac{\partial V(z, t)}{\partial t} \tag{6}$$

Equations (4) and (6) are known as the *telegrapher's equations* that provide a time-domain relationship between the voltage and current in any transmission line.

The above telegrapher's equations for voltage and current (Equations (4) and (6)) can be formalized in HOL Light in the time domain as follows:

Definition 4.1. *Telegrapher's Equation for Voltage*
$\vdash_{def}$ ∀V I L z t.
  telegraph_equation_voltage V I R L z t ⇔
    (complex_derivative (λz. V z t) z) =
      --(Cx L * complex_derivative (λt. I z t) t - Cx R * (I z t))

Definition 4.2. *Telegrapher's Equation for Current*
$\vdash_{def}$ ∀V I C z t.
  telegraph_equation_current V I G C z t ⇔
    (complex_derivative (λz. I z t) z) =
      --(Cx C * complex_derivative (λt. V z t) t) - Cx G * (V z t)

where `telegraph_equation_voltage` and `telegraph_equation_current` mainly accept the functions `V` and `I` of type $\mathbb{C} \times \mathbb{C} \to \mathbb{C}$, representing the voltage and current, respectively, and return the corresponding telegrapher's equations. The variables `R:`$\mathbb{R}$, `L:`$\mathbb{R}$, `G:`$\mathbb{R}$ `C:`$\mathbb{R}$, `z:`$\mathbb{C}$, and `t:`$\mathbb{C}$ represent the resistance, inductance, conductance, capacitance, the spatial coordinate and the time variable, respectively.

It is important to note that we use `complex_derivative` to formalize the time-domain PDEs due to the involvement of the phasor-domain representations of the voltage and current functions in the analysis. Since a phasor-domain representation of a function is a vector in complex plane with some magnitude and angle, the variables `z` and `t` are considered as complex numbers for convenience and the corresponding voltages and currents equations equally hold under this choice.

Now, we can combine the telegrapher's equations (Equations (4) and (6)) to obtain their alternate representations that are commonly known as the *wave equations*, which are more practical to use and provide some additional physical insights and are mathematically expressed as follows:

$$\frac{\partial^2 V(z,t)}{\partial z^2} - LC\frac{\partial^2 V(z,t)}{\partial t^2} = (RC + GL)\frac{\partial V(z,t)}{\partial t} + RGV(z,t) \qquad (7)$$

$$\frac{\partial^2 I}{\partial z^2} - LC\frac{\partial^2 I}{\partial t^2} = (RC + GL)\frac{\partial I(z,t)}{\partial t} + RGI(z,t) \qquad (8)$$

where $\frac{\partial^2}{\partial z^2}$ and $\frac{\partial^2}{\partial t^2}$ capture the second-order partial derivative with respect to z and t, respectively.

To model the wave equations for voltage and current, we need the transmission line constants, such as R, L, G and C. Therefore, we use the type abbreviation in HOL Light providing a compact representation of these constants as follows:

Definition 4.3. *Transmission Line Constants*
```
new_type_abbrev ("R",':ℝ')
new_type_abbrev ("L",':ℝ')
new_type_abbrev ("G",':ℝ')
new_type_abbrev ("C",':ℝ')
new_type_abbrev ("trans_line_const",':R # L # G # C')
```

Now, we formalize the wave equations for both voltage (Equation (7)) and current (Equation (8)) in the time domain as follows:

Definition 4.4. *Wave Equation for Voltage*
```
⊢_def ∀V R L G C z t.
wave_voltage_equation V ((R,L,G,C):trans_line_const) z t ⇔
 higher_complex_derivative 2 (λz. V z t) z -
  Cx L * Cx C * (higher_complex_derivative 2 (λt. V z t) t =
   (Cx R * Cx C + Cx G * Cx L) * (complex_derivative (λt. V z t) t) +
                              Cx R * Cx G * (V z t))
```

Definition 4.5. *Wave Equation for Current*
```
⊢_def ∀I R L G C z t.
 wave_current_equation I ((R,L,G,C):trans_line_const) z t ⇔
 higher_complex_derivative 2 (λz. I z t) z -
  Cx L * Cx C (higher_complex_derivative 2 (λt. I z t) t =
   (Cx R * Cx C + Cx G * Cx L) * (complex_derivative (λt. I z t) t) +
                              Cx R * Cx G * (I z t))
```

Next, we express the space-time voltage and current functions as *phasors* in order to reduce the PDEs to Ordinary Differential Equations (ODEs), which will greatly facilitate the derivation of the general solutions of these equations.

The relationship between the space-time voltage and current functions and their phasors can be mathematically expressed as follows [17]:

$$V(z,t) = \mathscr{R}e\{V(z)e^{j\omega t}\}$$

$$I(z,t) = \mathscr{R}e\{I(z)e^{j\omega t}\}$$

where $V(z)$ and $I(z)$ are the phasor components corresponding to $V(z,t)$ and $I(z,t)$, respectively.

4.2 Telegrapher's Equations in Phasor Domain

The principal advantage of the phasor representation of the telegrapher's equations over the time-domain versions is that we no longer need the derivatives with respect to time and are left with the derivatives with respect to distance only. This considerably simplifies the corresponding equations. For instance, the sinusoidally time-varying case, the telegrapher's equations (Equations (4) and (6)) can be rewritten in terms of phasor quantities by replacing $\dfrac{\partial}{\partial t}$ with $j\omega$. We can derive the telegrapher equation for voltage from Equation (4) as follows:

$$\frac{\partial V(z,t)}{\partial z} = -RI(z,t) - L\frac{\partial I(z,t)}{\partial t}$$

$$\frac{\partial}{\partial z}[\underbrace{\mathscr{R}e\{V(z)e^{j\omega t}\}}_{V(z,t)}] = -R[\underbrace{\mathscr{R}e\{I(z)e^{j\omega t}\}}_{I(z,t)}] - L\frac{\partial}{\partial t}[\underbrace{\mathscr{R}e\{I(z)e^{j\omega t}\}}_{I(z,t)}]$$

$$\mathscr{R}e\{e^{j\omega t}\frac{dV(z)}{dz}\} = \mathscr{R}e\{-RI(z)e^{jwt} - L(j\omega)e^{j\omega t}I(z)\}$$

$$\frac{dV(z)}{dz} = (-R - j\omega L)I(z)$$

From the above, we can rewrite the telegrapher's equations for voltage as:

$$\frac{dV(z)}{dz} + (R + j\omega L)I(z) = 0 \tag{9}$$

We can also derive the following Equation (10) from Equation (6) in a similar manner

$$\frac{dI(z)}{dz} + (G + j\omega C)V(z) = 0 \tag{10}$$

Here, Equations (9) and (10) are ODEs due to the fact that V and I are functions of the single variable z. Equation (9) indicates that the rate of change of the phasor voltage along the transmission line, as a function of position z, is equal to the series impedance of the line per unit length multiplied by the phasor current. Similarly, Equation (10) states that the rate of change of phasor current along the transmission line, as a function of position z, is equal to the shunt admittance of the line per unit length multiplied by the phasor voltage. We formalize the telegrapher's equation in the phasor domain for voltage (Equation (9)) as:

Definition 4.6. *Telegrapher's Equation*
$\vdash_{def}$ ∀V I R L w z. telegraph_equation_phasor_voltage V I R L w z ⇔
$\qquad\qquad\qquad\qquad\qquad\qquad$ telegraph_voltage V I R L w z = Cx(&0)

where `telegraph_equation_phasor_voltage` accepts the complex functions $V:\mathbb{C} \to \mathbb{C}$ and $I:\mathbb{C} \to \mathbb{C}$, the line parameters $R:\mathbb{R}$ and $L:\mathbb{R}$, the angular frequency $\omega:\mathbb{R}$, the spatial coordinate $z:\mathbb{C}$, and returns the corresponding telegrapher's equation. Here, the function `telegraph_voltage` models the left-hand side of Equation (9), and is formalized as follows:

Definition 4.7. *Left-Hand Side of Equation (9)*
$\vdash_{def}$ ∀V I R L w z.
 telegraph_voltage V I R L w z =
   complex_derivative (λz. V(z)) z + (Cx R + ii * Cx w * Cx L) * I(z)

Similarly, we formalize Equation (10) in HOL Light as follows:

Definition 4.8. *Telegrapher's Equation*
$\vdash_{def}$ ∀V I G C w z. telegraph_equation_phasor_current V I G C w z ⇔
$\qquad\qquad\qquad\qquad\qquad\qquad$ telegraph_current V I G C w z = Cx(&0)

with

Definition 4.9. *Left-Hand Side of Equation (10)*
$\vdash_{def}$ ∀V I G C w z. telegraph_current V I G C w z =
     complex_derivative (λz. I(z)) z + (Cx G + ii * Cx w * Cx C) * V(z)

where `telegraph_current` models the left-hand side of Equation (10).

4.3 Relationship between Telegrapher's and Wave Equations in Phasor Domain

A limitation in using the above form of the telegrapher's equations (Equations (9) and (10)) is that we need to solve each of them for both voltage and current.

To reduce such overhead, we can write the telegrapher's equations using one function ($V(z)$ or $I(z)$) as equivalent wave equations. To do this, we first take the derivative of Equation (9) with respect to z:

$$\frac{d}{dz}\left\{\frac{dV(z)}{dz} = -(R+j\omega L)I(z)\right\}$$

which can be written as:

$$\frac{d^2V(z)}{dz^2} = -(R+j\omega L)\frac{dI(z)}{dz} \tag{11}$$

Next, we substitute Equation (10) in Equation (11), to obtain the following equation that involves only $V(z)$:

$$\frac{d^2V(z)}{dz^2} = \gamma^2 V(z) \tag{12}$$

γ is the complex propagation constant and is mathematically expressed as:

$$\gamma = \alpha + j\beta = \sqrt{(R+j\omega L)(G+j\omega C)}. \tag{13}$$

where α is the attenuation coefficient and β is the phase coefficient and both are mathematically expressed as:

$$\alpha = \mathfrak{Re}(\gamma) = \mathfrak{Re}\{\sqrt{(R+j\omega L)(G+j\omega C)}\}$$

$$\beta = \mathfrak{Im}(\gamma) = \mathfrak{Im}\{\sqrt{(R+j\omega L)(G+j\omega C)}\}$$

In a similar manner, we derive the second wave equation by taking the derivative of Equation (10) and substituting Equation (9) in the resultant equation:

$$\frac{d^2I(z)}{dz^2} = \gamma^2 I(z) \tag{14}$$

We can alternatively represent the wave equations (Equations (12) and (14)) as:

$$\frac{d^2V(z)}{dz^2} - \gamma^2 V(z) = 0 \tag{15}$$

$$\frac{d^2I(z)}{dz^2} - \gamma^2 I(z) = 0 \tag{16}$$

Now, to verify a relationship between the telegrapher's and wave equations for voltage and current in the phasor domain, we first formalize the propagation constant in HOL Light as follows:

Definition 4.10. *Propagation Constant*

$\vdash_{def}$ ∀R L G C w.
```
propagation_constant ((R,L,G,C):trans_line_const) w =
    csqrt ((Cx R + ii * Cx w * Cx L) * (Cx G + ii * Cx w * Cx C))
```

The function `propagation_constant` accepts the transmission line parameters R, L, G, C and angular frequency ω, and returns the corresponding function.

The wave equations (Equations (15) and (16)) in higher-order-logic are formalized as:

Definition 4.11. *Wave Equation for Voltage*

$\vdash_{def}$ ∀V tlc w z.
```
  wave_equation_phasor_voltage V z tlc w ⇔
          wave_voltage V z tlc w = Cx(&0)
```

with

Definition 4.12. *Left-Hand Side of Equation (15)*

$\vdash_{def}$ ∀V tlc w z.
```
  wave_voltage V z tlc w =
      higher_complex_derivative 2 (λz. V(z)) z -
          (propagation_constant tlc w) pow 2 * V(z)
```

Definition 4.13. *Wave Equation for Current*

$\vdash_{def}$ ∀I tlc w z.
```
  wave_equation_phasor_current I z tlc w z ⇔
          wave_current I z tlc w = Cx(&0)
```

with

Definition 4.14. *Left-Hand Side of Equation (16)*

$\vdash_{def}$ ∀I tlc w z.
```
  wave_current I z tlc w =
      higher_complex_derivative 2 (λz. I(z)) z -
          (propagation_constant tlc w) pow 2 * I(z)
```

Now, we formally verify the relationship between the telegrapher's and wave equations for voltage in the phasor domain as the following HOL Light theorem:

Theorem 4.1. *Relationship between Telegrapher's and Wave Equations for Voltage*
$\vdash_{thm}$ ∀V I R L G C w z.
let tlc = ((R,L,G,C):trans_line_const) in
[A1] (λz. complex_derivative (λz. V z) z) complex_differentiable at z ∧
[A2] I complex_differentiable at z ∧
[A3] telegraph_current V I G C w z = Cx(&0)
  ⇒ complex_derivative (λz. telegraph_voltage V I R L w z) z =
                                        wave_voltage V z tlc w

Assumption **A1** ensures that the first-order derivative of function **V** is differentiable at **z**. Assumption **A2** asserts the differentiability of the function **I** at **z**. Assumption **A3** provides the telegrapher's equation for current, i.e., Equation (10). The proof of Theorem 4.1 is mainly based on the definitions of the telegrapher's and wave equations and some classical properties of the complex derivative along with some complex arithmetic reasoning. Similarly, we formally verify this relationship for current in the phasor domain.

Theorem 4.2. *Relationship between Telegrapher's and Wave Equations for Current*
$\vdash_{thm}$ ∀V I R L G C w z.
let tlc = ((R,L,G,C):trans_line_const) in
[A1] (λz. complex_derivative (λz. I z) z) complex_differentiable at z ∧
[A2] V complex_differentiable at z ∧
[A3] telegraph_voltage V I R L w z = Cx(&0)
  ⇒ complex_derivative (λz. telegraph_current V I G C w z) z =
                                        wave_current I z tlc w

The verification of Theorem 4.2 is very similar to that of Theorem 4.1. More details about their verification can be found at [10].

5 Formal Verification of Analytical Solutions of the Telegrapher's Equations

Analyzing transmission lines is mainly based on finding out solutions of these PDE based telegrapher's and wave equations that are further used to analyze various aspects of signal propagation, such as attenuation, distortion, reflection, and dispersion along the transmission line. One of the examples is to understand the behavior of high-frequency signals, where the distributed parameters of the transmission line significantly affect the signal integrity. In this section, we formally verify the correctness of the analytical solutions of the telegrapher's equations in the phasor domain pertaining to sinusoidal steady state and in the time domain that are concerned with arbitrary variations over time.

5.1 Verification of the Solutions in Phasor Domain

We can mathematically express the general solutions of the wave equations (and thus the telegrapher's equations) (Equations (15) and (16)) as folllows:

$$V(z) = V^+(z) + V^-(z) = V_0^+ e^{-\gamma z} + V_0^- e^{\gamma z} \tag{17}$$

$$I(z) = I^+(z) + I^-(z) = I_0^+ e^{-\gamma z} + I_0^- e^{\gamma z} \tag{18}$$

Here, V_0^+, V_0^-, I_0^+, I_0^- are the complex constants that can be determined by boundary conditions. Similarly, the transmission line voltage $V^+(z)$ and current $I^+(z)$ represent the forward-going waves (propagating in the $+z$ direction) and voltage $V^-(z)$ and current $I^-(z)$ are the backward-going waves (propagating in the $-z$ direction).

If we insert the solution for $V(z)$ in Equation (9), we get:

$$\frac{dV(z)}{dz} = -\gamma V_0^+ e^{-\gamma z} + \gamma V_0^- e^{\gamma z} = -(R + j\omega L)I(z) \tag{19}$$

Next, we rearrange the above equation to obtain the current $I(z)$:

$$I(z) = \frac{\gamma}{R + j\omega L}(V_0^+ e^{-\gamma z} - V_0^- e^{\gamma z}) \tag{20}$$

Note that both expressions (Equations (18) and (20)) for the current are the same. The characteristic impedance, which is the ratio of the line voltage and current, is an important characteristic of transmission line and can be mathematically expressed as follows:

$$Z_0 = \frac{V_0^+}{I_0^+} = \frac{-V_0^-}{I_0^-} = \sqrt{\frac{R + j\omega L}{G + j\omega C}} = \frac{R + J\omega L}{\gamma} = R_0 + jX_0 \tag{21}$$

where R_0 and X_0 are the real and imaginary parts of Z_0. The characteristic impedance Z_0 and the propagation constant γ are two important properties of the transmission line due to their direct dependence on the line parameters R, L, G, C and the phasor of the operation.

Next, we define the characteristic impedance in HOL Light as follows:

Definition 5.1. *Characteristic Impedance*

$\vdash_{def} \forall$R L G C w.

```
  characteristic_impedance (R,L,G,C) w =
    (let tlc = ((R,L,G,C):trans_line_const) in
      (Cx R + ii * Cx w * Cx L) / propagation_constant tlc w)
```

The next step is to formalize the general solutions (Equations (17) and (20)) in HOL Light:

Definition 5.2. *Wave Solution for Voltage*
$\vdash_{def}$ ∀V1 V2 tlc w z.
    wave_solution_voltage_phasor V1 V2 tlc w z =
      V1 * cexp(--(propagation_constant tlc w) * z) +
       V2 * cexp((propagation_constant tlc w) * z)

where V1 and V2 in the formalization refer to the complex constants V_0^+ and V_0^- in Equation (17), respectively. The parameters w and z represent the angular frequency and the spatial coordinate, respectively.

Definition 5.3. *Wave Solution for Current*
$\vdash_{def}$ ∀V1 V2 tlc w z.
    wave_solution_current_phasor V1 V2 tlc w z =
    Cx(&1) / characteristic_impedance tlc w *
      (V1 * cexp(--(propagation_constant tlc w) * z) -
       V2 * cexp((propagation_constant tlc w) * z))

Next, we formally verify the general solutions (Equations (17) and (20)) of the wave equations for voltage and current, (represented by Equations (15) and (16)), in the HOL Light theorem prover as follows:

Theorem 5.1. *Correctness of the Solution for Voltage*
$\vdash_{thm}$ ∀V1 V2 V R L G C w z.
    let tlc = ((R,L,G,C):trans_line_const) in
      wave_equation_voltage_phasor
        (λz. wave_solution_voltage_phasor V1 V2 tlc w z) V tlc w

Theorem 5.2. *Correctness of the Solution for Current*
$\vdash_{thm}$ ∀V1 V2 I R L G C w z.
    let tlc = ((R,L,G,C):trans_line_const) in
      wave_equation_current_phasor
        (λz. wave_solution_current_phasor V1 V2 tlc w z) I tlc w

The verification of Theorems 5.1 and 5.2 is mainly based on four lemmas about the complex differentiation of the solutions, given in Table 3.

Mathematical Form	Formalized Form

$\dfrac{dV(z)}{dz} = -\gamma V1e^{-\gamma z} + \gamma V2e^{\gamma z}$	**Lemma 1 (First-Order Derivative of General Solution for Voltage):**
	∀V1 V2 R L G C w z.
	  let tlc = ((R,L,G,C):trans_line_const) in
	    complex_derivative (λz.
	      wave_solution_voltage_phasor V1 V2 tlc w z) z =
	      V1 * (--(propagation_constant tlc w)) *
	        cexp (--(propagation_constant tlc w) * z) +
	          V2 * (propagation_constant tlc w) *
	            cexp ((propagation_constant tlc w) * z)

$\dfrac{d^2V(z)}{dz^2} = \gamma^2 V1e^{-\gamma z} + \gamma^2 V2e^{\gamma z}$	**Lemma 2 (Second-Order Derivative of General Solution for Voltage):**
	∀V1 V2 R L G C w z.
	  let tlc = ((R,L,G,C):trans_line_const) in
	    higher_complex_derivative 2 (λz.
	      wave_solution_voltage_phasor V1 V2 tlc w z) z =
	      V1 * (propagation_constant tlc w) pow 2 *
	        cexp (--(propagation_constant tlc w) * z) +
	          V2 * (propagation_constant tlc w) pow 2 *
	            cexp ((propagation_constant tlc w) * z

$\dfrac{dI(z)}{dz} = \dfrac{1}{Z_0}(-\gamma V1e^{-\gamma z} - \gamma V2e^{\gamma z})$	**Lemma 3 (First-Order Derivative of General Solution for Current):**
	∀V1 V2 R L G C w z.
	  let tlc = ((R,L,G,C):trans_line_const) in
	    complex_derivative (λz.
	      wave_solution_current_phasor V1 V2 tlc w z) z =
	      Cx (&1) / characteristic_impedance tlc w *
	       (V1 * (--propagation_constant tlc w) *
	         cexp (--(propagation_constant tlc w) * z) -
	          V2 * (propagation_constant tlc w) *
	           cexp ((propagation_constant tlc w) * z))

$\dfrac{d^2I(z)}{dz^2} = \dfrac{1}{Z_0}(\gamma^2 V1e^{-\gamma z} - \gamma^2 V2e^{\gamma z})$	**Lemma 4 (Second-Order Derivative of General Solution for Current):**
	∀V1 V2 R L G C w z.
	  let tlc = ((R,L,G,C):trans_line_const) in
	    higher_complex_derivative 2 (λz.
	      wave_solution_current_phasor V1 V2 tlc w z) z =
	      Cx (&1) / characteristic_impedance tlc w *
	       (V1 * (propagation_constant tlc w) pow 2 *
	         cexp (--(propagation_constant tlc w) * z) -
	          V2 * (propagation_constant tlc w) pow 2 *
	           cexp ((propagation_constant tlc w) * z))

Table 3: Lemmas of the Derivatives of General Solutions in Phasor Domain

Since, there exists a relationship between the telegrapher's and wave equations, as proven in Section 4, the solutions of the wave equations also satisfy the telegrapher's equations.

215

Theorem 5.3. *General Solution of the Telegrapher's Equation for Voltage*
```
⊢_thm ∀V1 V2 V I R L G C w.
 let tlc = ((R,L,G,C):trans_line_const) in
[A1] Cx R + ii * Cx w * Cx L ≠ Cx(&0) ∧
[A2] (∀z. V z = wave_solution_voltage_phasor V1 V2 tlc w z) ∧
[A3] (∀z. I z = wave_solution_current_phasor V1 V2 tlc w z)

 ⇒ telegraph_equation_phasor_voltage V I R L w z
```

Assumption A1 ensures that expression $R + j\omega L$ is not equal to zero. Assumptions A2 and A3 provide solutions of the wave equations for the voltage and the current, respectively. The verification of the above theorem is based on the properties of the complex differentiation along with some complex arithmetic reasoning.

Theorem 5.4. *General Solution of the Telegrapher's Equation for Current*
```
⊢_thm ∀V1 V2 V I R L G C w.
 let tlc = ((R,L,G,C):trans_line_const) in
[A1] Cx R + ii * Cx w * Cx L ≠ Cx(&0) ∧
[A2] (∀z. V z = wave_solution_voltage_phasor V1 V2 tlc w z) ∧
[A3] (∀z. I z = wave_solution_current_phasor V1 V2 tlc w z)

 ⇒ telegraph_equation_phasor_current V I G C w z
```

5.2   Verification of Properties of Transmission Line

A transmission line is characterized by two essential properties, namely its propagation constant γ and characteristic impedance Z_0. These properties are specified by the angular frequency ω and the line parameters R, L, G and C. Understanding and optimizing the transmission line characteristics help engineers and designers to achieve efficient signal transmission, maintain signal integrity, and ensure the reliable operation of these systems. In this section, we formally verify these transmission line properties for the case of lossless and distortionless lines.

5.2.1   Lossless Line

The main purpose of a transmission line is to facilitate the transmission of information between distant locations with minimal signal degradation that can be achieved by reducing the signal loss. This is one of the crucial requirements in the construction of an efficient and a reliable transmission line. In the case of a lossless transmission line, the elements R (resistance) and G (conductance) can be considered as negligible or effectively zero:

$$R = G = 0$$

The characteristic impedance of a lossless transmission line can now be expressed in a simplified form by using the above values of R and G in Equation (21) as:

$$Z_0 = \sqrt{\frac{j\omega L}{j\omega C}} = \sqrt{\frac{L}{C}}$$

Similarly, the attenuation and phase constants expressed in Equation (13) becomes:

$$\alpha = 0 \tag{22}$$

$$\beta = \sqrt{LC} \tag{23}$$

This implies that the transmission line has no signal attenuation, and as a result, the propagation constant can be represented by a purely imaginary number:

$$\gamma = j\beta = j\omega\sqrt{LC}$$

5.2.2 Distortionless Line

A distortionless line refers to a transmission medium characterized by an attenuation constant α that exhibits no variation with changes in frequency while the phase constant β is linearly dependent on frequency.

For a distortionless transmission line, the elements R and G are related as:

$$\frac{R}{L} = \frac{G}{C}$$

Now, the characteristic impedance of the transmission line is expressed as:

$$Z_0 = \sqrt{\frac{R(1 + j\omega L/R)}{R(1 + j\omega C/G)}} = \sqrt{\frac{R}{G}} = \sqrt{\frac{L}{C}}$$

The propagation constant (Equation (13)) becomes:

$$\gamma = \sqrt{RG\left(1 + \frac{j\omega L}{R}\right)\left(1 + \frac{j\omega C}{G}\right)}$$

$$\gamma = \sqrt{RG}\left(1 + \frac{j\omega C}{G}\right) = \alpha + j\beta$$

or

$$\alpha = \sqrt{RG}, \qquad \beta = \omega\sqrt{LC} \qquad (24)$$

We can see that the attenuation constant α is independent of the frequency, whereas β is a linear function of frequency.

The verified properties, i.e., propagation constant and characteristic impedance of the lossless and distortionless transmission lines are given in Table 4.

Case	Propagation Constant	Characteristic Impedance
Lossless	**Theorem 1 (*Attenuation Constant*)** $\vdash_{thm} \forall$R L G C w. let tlc = ((R,L,G,C):trans_line_const) in [A1] w > &0 [A2] L > &0 $\wedge$ [A3] C > &0 $\wedge$ [A4] R = &0 $\wedge$ [A5] G = &0 $\Rightarrow$ Re(propagation_constant tlc w) = &0 **Theorem 2 (*Phase Constant*)** $\vdash_{thm} \forall$R L G C w. let tlc = ((R,L,G,C):trans_line_const) in [A1] w > &0 [A2] L > &0 $\wedge$ [A3] C > &0 $\wedge$ [A4] R = &0 $\wedge$ [A5] G = 0 $\Rightarrow$ Im(propagation_constant tlc w) = w$\sqrt{LC}$	**Theorem 3 (*Characteristic Impedance*)** $\vdash_{thm} \forall$R L G C w. let tlc = ((R,L,G,C):trans_line_const) in [A1] w > &0 [A2] L > &0 $\wedge$ [A3] C > &0 $\wedge$ [A4] R = &0 $\wedge$ [A5] G = &0 [A6] ii * Cx w $\neq$ Cx (&0) [A7] csqrt (Cx L * Cx C) $\neq$ Cx (&0) $\Rightarrow$ characteristic_impedance tlc w = csqrt(Cx(L) * Cx(C)) / Cx(C)
Distortionless	**Theorem 4 (*Attenuation Constant*)** $\vdash_{thm} \forall$R L G C w. let tlc = ((R,L,G,C):trans_line_const) in [A1] L > &0 $\wedge$ [A2] R > &0 $\wedge$ [A3] G > &0 $\wedge$ [A4] R / L = G / C $\Rightarrow$ Re(propagation_constant tlc w) = $\sqrt{RG}$ **Theorem 5 (*Phase Constant*)** $\vdash_{thm} \forall$R L G C w. let tlc = ((R,L,G,C):trans_line_const) in [A1] L > &0 $\wedge$ [A2] R > &0 $\wedge$ [A3] G > &0 $\wedge$ [A4] C > &0 $\wedge$ [A5] R / L = G / C $\Rightarrow$ Im(propagation_constant tlc w) = w$\sqrt{LC}$	**Theorem 6 (*Characteristic Impedance*)** $\vdash_{thm} \forall$R L G C w. let tlc = ((R,L,G,C):trans_line_const) in [A1] L > &0 $\wedge$ [A2] C > &0 $\wedge$ [A3] &0 < R $\wedge$ [A4] G > &0 $\wedge$ [A5] Cx G + ii * Cx w * Cx C $\neq$ Cx (&0) $\wedge$ [A6] R / L = G / C $\Rightarrow$ characteristic_impedance tlc w = csqrt(Cx(L) * Cx(C)) / Cx(C)

Table 4: Properties of Transmission Lines

In the following section, we verify the general solutions of the time-domain PDEs by considering a lossless line, where we assume both resistance R and conductance G to be zero.

5.3   Verification of the Solutions in Time Domain

It is useful to examine the complete time functions for understanding the nature of the voltage and current within a transmission line. We can find the corresponding time-domain expressions for voltage and current (solution in the time domain) on the line by multiplying the phasor of the voltage and current with the harmonic time variation term e^{jwl} and taking its real part as follows:

$$V(z,t) = \mathscr{R}e\{V(z)e^{j\omega t}\} \tag{25}$$

$$I(z,t) = \mathscr{R}e\{I(z)e^{j\omega t}\} \tag{26}$$

Next, we use Equation (17) in the time-domain solution (Equation (25)) and get:

$$V(z,t) = \mathscr{R}e\{(V_0^+ e^{-\gamma z} + V_0^- e^{\gamma z})e^{j\omega t}\}$$

$$V(z,t) = \mathscr{R}e\{V_0^+ e^{-\gamma z} e^{j\omega t} + V_0^- e^{\gamma z} e^{j\omega t}\}$$

By splitting the propagation constant in real and imaginary parts, i.e., $\gamma = \alpha + j\beta$, we can write the above equation for voltage as follows:

$$V(z,t) = \mathscr{R}e\{V_0^+ e^{-(\alpha+j\beta)z} e^{j\omega t} + V_0^- e^{(\alpha+j\beta)z} e^{j\omega t}\}$$

We know that α is equal to zero for a lossless transmission line. Thus, we get:

$$V(z,t) = \mathscr{R}e\{V_0^+ e^{j(\omega t - \beta z)} + V_0^- e^{j(\omega t + \beta z)}\} \tag{27}$$

After applying Euler's formula to the above equation and taking the real part of the solution, we have:

$$V(z,t) = V_0^+ \cos(\omega t - \beta z) + V_0^- \cos(\omega t + \beta z) \tag{28}$$

where we assume V_0^+ and V_0^- to be real.

Using Definition 5.2, we formalize the general solution (Equation (25)) in the time-domain for voltage as follows:

Definition 5.4. *General Solution for Voltage in Time Domain*
$\vdash_{def}$ ∀V1 V2 tlc w z t.
  wave_solution_voltage_time V1 V2 tlc w z t =
    Re((wave_solution_voltage_phasor V1 V2 tlc w z) * cexp(ii * Cx w * t))

where the function `wave_solution_voltage_time` uses the phasor given by the voltage function `wave_solution_voltage_phasor` to construct the formal definition of Equation (25).

Next, we formally verify the general solution for voltage in the time domain in HOL Light as follows:

Theorem 5.5. *General Solution of Wave Equation for Voltage*
$\vdash_{thm}$ ∀V1 V2 R L G C w.

```
let tlc = ((R,L,G,C):trans_line_const) in
[A1]  w > &0 ∧ [A2]  L > &0 ∧ [A3]  C > &0 ∧
[A4]  R = &0 ∧ [A5]  G = &0 ∧ [A6]  (∀t. Im t = &0) ∧
[A7]  (∀z. Im z = &0) ∧ [A8]  Im V1 = &0 ∧ [A9]  Im V2 = &0 ∧
[A10](∀z t. V z t = Cx(wave_solution_voltage_time V1 V2 tlc w z t))
     ⇒ wave_voltage_equation V tlc z t
```

Assumptions **A1–A3** ensure that the angular frequency ω, the line parameters **L** and **C** are positive real values. Assumptions **A4–A5** assert that the line parameters **R** and **G** are equal to zero, which is an assumption for a lossless transmission line. Assumptions **A6–A7** ensure that the imaginary parts of the variables **z** and **t** are equal to zero in the time domain. Assumptions **A8–A9** guarantee that the coefficients **V1** and **V2** are real. Assumption **A10** provides the solution of the wave equation for voltage, i.e., Equation (28). The proof of the above theorem is mainly based on the following Lemma 5.1 which gives the relationship between phasor and time-domain functions as well as four important lemmas about the complex differentiation of the time-domain solution with respect to the parameters **z** and **t**, which are given in Table 5.

Lemma 5.1. *Relationship between Phasor and Time-Domain Functions for Voltage*
$\vdash_{lem}$ ∀V1 V2 R C L G w z t.

```
let tlc = ((R,L,G,C):trans_line_const) in
[A1]  w > &0 ∧ [A2]  L > &0 ∧ [A3]  C > &0 ∧
[A4]  R = &0 ∧ [A5]  G = &0
  ⇒ wave_solution_voltage_time V1 V2 tlc w z t =
     Re(V1) * (cos(w * Re t - (Im(propagation_constant tlc w)) * Re z)) +
       Re(V2) * (cos(w * Re t + Im((propagation_constant tlc w)) * Re z))
```

Assumptions **A1–A5** are the same as those of Theorem 5.5. The verification of Lemma 5.1 is mainly based on Theorem 1 given in Table 4 and the properties of transcendental functions alongside some complex arithmetic reasoning.

220

Mathematical Form	Formalized Form

$\dfrac{\partial V(z,t)}{\partial z} = V1sin(\omega t - \beta z)\beta - V2sin(\omega t + \beta z)\beta$	**Lemma 1**
	(First-Order Partial Derivative of General Solution
	       for Voltage with respect to distance):
	∀V1 V2 R L G C w.
	let tlc = ((R,L,G,C):trans_line_const)
	[A1] w > &0 ∧ **[A2]** L > &0 ∧ **[A3]** C > &0 ∧
	[A4] R = &0 **[A5]** G = &0 ∧ **[A6]** (∀t. Im t = &0) ∧
	[A7] (∀z. Im z = &0) ∧ **[A8]** Im V1 = &0 ∧ **[A9]** Im V2 = &0 ∧
	⇒ complex_derivative (λz.
	    wave_solution_voltage_time V1 V2 tlc w z t) z =
	      Cx (Re V1 * (--sin (w * Re t - (w * sqrt (L * C)) * Re z))·
	        (--(w * sqrt (L * C)))) + Re V2 * (--sin (w * Re t +
	          (w * sqrt (L * C)) * Re z))* ((w * sqrt (L * C)))

$\dfrac{\partial^2 V(z,t)}{\partial z^2} = -V1cos(\omega t - \beta z)\beta^2 - V2cos(\omega t + \beta z)\beta^2$	**Lemma 2**
	(Second-Order Partial Derivative of General Solution
	       for Voltage with respect to distance):
	∀V1 V2 R L G C w.
	let tlc = ((R,L,G,C):trans_line_const)
	[A1] w > &0 ∧ **[A2]** L > &0 ∧ **[A3]** C > &0 ∧
	[A4] R = &0 **[A5]** G = &0 ∧ **[A6]** (∀t. Im t = &0) ∧
	[A7] (∀z. Im z = &0) ∧ **[A8]** Im V1 = &0 ∧ **[A9]** Im V2 = &0 ∧
	⇒ higher_complex_derivative 2 (λz.
	    wave_solution_voltage_time V1 V2 tlc w z t) z =
	      Cx (Re V1 * (--cos (w * Re t - (w * sqrt (L * C)) * Re z))·
	        ((w * sqrt (L * C))) pow 2 + Re V2 * (--cos (w * Re t +
	          (w * sqrt (L * C)) * Re z)) * ((w * sqrt (L * C))) pow 2)

$\dfrac{\partial V(z,t)}{\partial t} = -V1sin(\omega t - \beta z)\omega - V2sin(\omega t + \beta z)\omega$	**Lemma 3**
	(First-Order Partial Derivative of General Solution
	       for Voltage with respect to time):
	∀V1 V2 R L G C w.
	let tlc = ((R,L,G,C):trans_line_const)
	[A1] w > &0 ∧ **[A2]** L > &0 ∧ **[A3]** C > &0 ∧
	[A4] R = &0 **[A5]** G = &0 ∧ **[A6]** (∀t. Im t = &0) ∧
	[A7] (∀z. Im z = &0) ∧ **[A8]** Im V1 = &0 ∧ **[A9]** Im V2 = &0 ∧
	⇒ complex_derivative (λt.
	    wave_solution_voltage_time V1 V2 tlc w z t) t =
	      Cx (Re V1 * (--sin (w * Re t - (w * sqrt (L * C)) * Re z)) * w +
	        Re V2 * (--sin (w * Re t + (w * sqrt (L * C)) * Re z)) * w)

$\dfrac{\partial^2 V(z,t)}{\partial t^2} = -V1cos(\omega t - \beta z)\omega^2 - V2cos(\omega t + \beta z)\omega^2$	**Lemma 4**
	(Second-Order Partial Derivative of General Solution
	       for Voltage with respect to time):
	∀V1 V2 R C L G w.
	let tlc = ((R,L,G,C):trans_line_const)
	[A1] w > &0 ∧ **[A2]** L > &0 ∧ **[A3]** C > &0 ∧
	[A4] R = &0 **[A5]** G = &0 ∧ **[A6]** (∀t. Im t = &0) ∧
	[A7] (∀z. Im z = &0) ∧ **[A8]** Im V1 = &0 ∧ **[A9]** Im V2 = &0 ∧
	⇒ higher_complex_derivative 2 (λt.
	    wave_solution_voltage_time V1 V2 tlc w z t) t =
	      Cx (Re V1 * (--cos (w * Re t - (w * sqrt (L * C)) * Re z)) * w pow 2 +
	        Re V2 * (--cos (w * Re t + (w * sqrt (L * C)) * Re z)) * w pow 2)

Table 5: Lemmas of the Derivatives of General Solutions for Voltage in Time Domain

221

Similarly, we use Equation (20) in the time domain solution (Equation (26)) for current as follows:

$$I(z,t) = \mathscr{R}e\{\frac{\gamma}{R+j\omega L}(V_0^+ e^{-\gamma z} - V_0^- e^{\gamma z})e^{j\omega t}\}$$

After rearranging the above equation, we have:

$$I(z,t) = \mathscr{R}e\{\frac{\gamma}{R+j\omega L}(V_0^+ e^{j(\omega t-\beta z)} - V_0^- e^{j(\omega t+\beta z)})\} \tag{29}$$

Next, by applying Euler's formula and taking the real part of the solution, we get:

$$I(z,t) = \frac{\gamma}{R+j\omega L}(V_0^+ \cos(\omega t - \beta z) - V_0^- \cos(\omega t + \beta z)) \tag{30}$$

Now, using Definition 5.3, we formalize the general solution (Equation (26)) in the time domain for current as follows:

Definition 5.5. *General Solution for Current in Time Domain*
$\vdash_{def} \forall$V1 V2 tlc w z t.
  wave_solution_current_time V1 V2 tlc w z t =
    Re((wave_solution_current_phasor V1 V2 tlc w z) * cexp(ii * Cx w * t))

where wave_solution_current_time accepts the phasor solution of the current wave_solution _current_phasor that is multiplied with the harmonic time variation term and returns its real part.

Theorem 5.6. *General Solution of Wave Equation for Current*
$\vdash_{thm} \forall$V1 V2 R L G C w.
  let tlc = ((R,L,G,C):trans_line_const) in
  [A1] w > &0 $\wedge$ [A2] L > &0 $\wedge$ [A3] C > &0 $\wedge$
  [A4] R = &0 $\wedge$ [A5] G = &0 $\wedge$ [A6] ($\forall$t. Im t = &0) $\wedge$
  [A7] ($\forall$z. Im z = &0) $\wedge$ [A8] Im V1 = &0 $\wedge$ [A9] Im V2 = &0 $\wedge$
  [A10] Im(Cx(&1)/characteristic_impedance tlc w) = &0 $\wedge$
  [A11] ($\forall$z t. I z t = Cx(wave_solution_current_time V1 V2 tlc w z t))
      $\Rightarrow$ wave_current_equation I tlc z t

Assumptions A1–A9 are the same as those of Theorem 5.5. Assumption A10 ensures that the imaginary part of the inverse characteristic impedance is equal to zero. Assumption A11 provides the solution of the wave equation for current, i.e., Equation (30). Similarly, the proof of Theorem 5.6 is primarily based on the formally verified lemmas about the relationship between phasor and time-domain functions, i.e., Lemma 5.2 and derivatives of the general solution for current as given in Table 6.

222

Mathematical Form	Formalized Form

$$\frac{\partial I(z,t)}{\partial z} = \frac{1}{Z_0}(V1sin(\omega t - \beta z)\beta + \\ V2sin(\omega t + \beta z)\beta)$$	**Lemma 1**
	(First-Order Partial Derivative of General Solution
	      for Current with respect to distance):
	∀V1 V2 R L G C w.
	let tlc = ((R,L,G,C):trans_line_const) in
	[A1] w > &0 ∧ [A2] L > &0 ∧ [A3] C > &0 ∧
	[A4] R = &0 ∧ [A5] G = &0 ∧ [A6] (∀t. Im t = &0) ∧
	[A7] (∀z. Im z = &0) ∧ [A8] Im V1 = &0 ∧ [A9] Im V2 = &0
	⇒ complex_derivative (λz.
	    wave_solution_current_time V1 V2 tlc w z t) z =
	      Cx (Re ((Cx (&1) / characteristic_impedance tlc w)) *
	      (Re V1 * --sin (w * Re t - (w * sqrt (L * C)) * Re z) *
	      --(w * sqrt (L * C)) + Re V2 * sin (w * Re t +
	      (w * sqrt (L * C)) * Re z) * (w * sqrt (L * C))))

$$\frac{\partial^2 I(z,t)}{\partial z^2} = \frac{1}{Z_0}(-V1cos(\omega t - \beta z)\beta^2 + \\ V2cos(\omega t + \beta z)\beta^2)$$	**Lemma 2**
	(Second-Order Partial Derivative of General Solution
	      for Current with respect to distance):
	∀V1 V2 R L G C w.
	let tlc = ((R,L,G,C):trans_line_const) in
	[A1] w > &0 ∧ [A2] L > &0 ∧ [A3] C > &0 ∧
	[A4] R = &0 [A5] G = &0 ∧ [A6] (∀t. Im t = &0) ∧
	[A7] (∀z. Im z = &0) ∧ [A8] Im V1 = &0 ∧ [A9] Im V2 = &0 ∧
	⇒ higher_complex_derivative 2 (λz.
	    wave_solution_current_time V1 V2 tlc w z t) z =
	      Cx (Re (Cx (&1) / characteristic_impedance tlc w) *
	      (Re V1 * --cos (w * Re t - (w * sqrt (L * C)) * Re z) *
	      (w * sqrt (L * C)) pow 2 + Re V2 * cos (w * Re t +
	      (w * sqrt (L * C)) * Re z) * (w * sqrt (L * C)) pow 2))

$$\frac{\partial I(z,t)}{\partial t} = \frac{1}{Z_0}(-V1sin(\omega t - \beta z)\omega + \\ V2sin(\omega t + \beta z)\omega)$$	**Lemma 3**
	(First-Order Partial Derivative of General Solution
	      for Current with respect to time):
	∀V1 V2 R L G C w.
	let tlc = ((R,L,G,C):trans_line_const) in
	[A1] w > &0 ∧ [A2] L > &0 ∧ [A3] C > &0 ∧
	[A4] R = &0 [A5] G = &0 ∧ [A6] (∀t. Im t = &0) ∧
	[A7] (∀z. Im z = &0) ∧ [A8] Im V1 = &0 ∧ [A9] Im V2 = &0 ∧
	⇒ complex_derivative (λt.
	    wave_solution_current_time V1 V2 tlc w z t) t =
	      Cx (Re (Cx (&1) / characteristic_impedance tlc w) *
	      Cx (Re V1 * (--sin (w * Re t - (w * sqrt (L * C)) *
	      Re z)) * w + Re V2 * (sin (w * Re t +
	      (w * sqrt (L * C)) * Re z)) * w)

$$\frac{\partial^2 I(z,t)}{\partial t^2} = \frac{1}{Z_0}(-V1cos(\omega t - \beta z)\omega^2 + \\ V2cos(\omega t + \beta z)\omega^2)$$	**Lemma 4**
	(Second-Order Partial Derivative of General Solution
	      for Current with respect to time):
	∀V1 V2 R L G C w.
	let tlc = ((R,L,G,C):trans_line_const) in
	[A1] w > &0 ∧ [A2] L > &0 ∧ [A3] C > &0 ∧
	[A4] R = &0 [A5] G = &0 ∧ [A6] (∀t. Im t = &0) ∧
	[A7] (∀z. Im z = &0) ∧ [A8] Im V1 = &0 ∧ [A9] Im V2 = &0 ∧
	⇒ higher_complex_derivative 2 (λt.
	wave_solution_current_time V1 V2 tlc w z t) t =
	Cx (Re (Cx (&1) / characteristic_impedance tlc w) * (Re V1 *
	  --cos (w * Re t - (w * sqrt (L * C)) * Re z) *
	  (w * sqrt (L * C)) pow 2 + Re V2 * cos (w * Re t +
	  (w * sqrt (L * C)) * Re z) * (w * sqrt (L * C)) pow 2))

Table 6: Lemmas of the Derivatives of General Solutions for Current in Time Domain

Lemma 5.2. *Relationship between Phasor and Time-Domain Functions for Current*
⊢_lem ∀V1 V2 R L G C w z t.
  let tlc = ((R,L,G,C):trans_line_const) in
  [A1] w > &0 ∧ [A2] L > &0 ∧ [A3] C > &0 ∧
  [A4] R = &0 ∧ [A5] G = &0
    ⇒ wave_solution_current_time V1 V2 tlc w z t =
      Re(Cx(&1) / characteristic_impedance tlc w) *
        (Re V1 * cos(w * Re t - Im(propagation_constant tlc w) * Re z) -
          Re V2 * cos(w * Re t + Im(propagation_constant tlc w) * Re z))

Assumptions A1–A5 are the same as those of Lemma 5.1. The verification of the above lemma is similar to that of Lemma 5.1.

Since the wave and telegrapher's equations are related to each other, the general solutions of the wave equations satisfy the telegrapher's equations in the time domain and are verified as the following HOL Light theorems:

Theorem 5.7. *General Solution of Telegrapher's Equation for Voltage*
⊢_thm ∀V1 V2 R L G C w.
  let tlc = ((R,L,G,C):trans_line_const) in
  [A1] w > &0 ∧ [A2] L > &0 ∧ [A3] C > &0 ∧
  [A4] R = &0 ∧ [A5] G = &0 ∧ [A7] (∀t. Im t = &0) ∧
  [A8] (∀z. Im z = &0) ∧ [A9] Im V1 = &0 ∧ [A10] Im V2 = &0 ∧
  [A11] (∀z t. V z t = Cx (wave_solution_voltage_time V1 V2 tlc w z t))
    ⇒ telegraph_equation_voltage V I R L z t

Theorem 5.8. *General Solution of Telegrapher's Equation for Current*
⊢_thm ∀V1 V2 R L G C w.
  let tlc = ((R,L,G,C):trans_line_const) in
  [A1] w > &0 ∧ [A2] L > &0 ∧ [A3] C > &0 ∧
  [A4] R = &0 ∧ [A5] G = &0 ∧ [A6] (∀t. Im t = &0) ∧
  [A7] (∀z. Im z = &0) ∧ [A8] Im V1 = &0 ∧ [A9] Im V2 = &0 ∧
  [A10] Im (Cx(&1)/characteristic_impedance tlc w) = &0 ∧
  [A11] (∀z t. I z t = Cx (wave_solution_current_time V1 V2 tlc w z t))
    ⇒ telegraph_equation_current V I G C z t

Assumptions of the above theorems are the same as those of Theorems 5.5 and 5.6. Similar to the verification of the wave equations in the time domain, we used Lemmas 5.1 and 5.2 as well as the verified lemmas of the derivatives for voltage and current in order to verify the correctness of the wave solutions for the telegrapher's equations. More details about the verification of the time-domain PDEs can be found in our proof script [10].

6 Application: Terminated Transmission Line

To illustrate the practical effectiveness of our proposed approach, we formally analyze the behavior of various transmission lines connected between a generator and a load. Particularly, we perform a formal analysis of a terminated transmission line by formally verifying the load impedance and the voltage reflection coefficient. Moreover, we formally analyze short-circuited and open-circuited transmission lines that are commonly used in the construction of resonant circuits and matching stubs. These lines correspond to the special cases of the load impedance: $Z_L = 0$ for a short-circuited line and $Z_L = \infty$ for an open-circuited line.

Terminated transmission lines in arbitrary complex load impedances are used in the majority of sinusoidal steady-state applications. They play a vital role in ensuring a smooth transfer of signals or power, especially in applications where signal quality and system performance are critical. We consider the essential behavior of line voltage, current, and impedance for a portion of a lossless transmission line terminated with a load Z_L, as shown in Figure 3. In this section, we formally analyze a terminated transmission line by formally verifying in HOL Light various important properties, such as load impedance and voltage reflection coefficient.

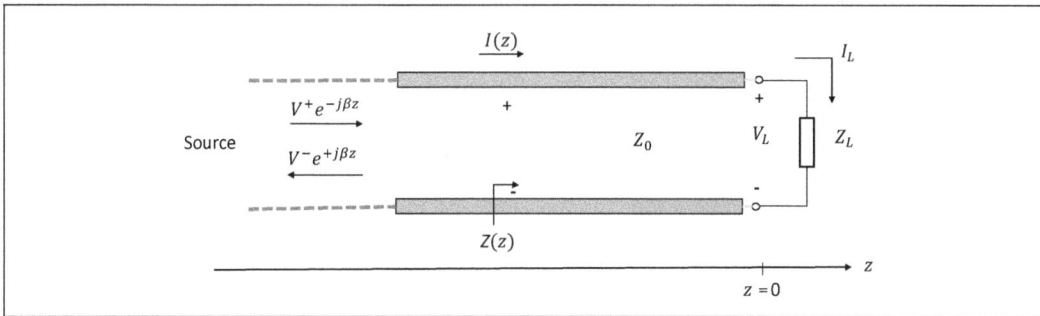

Figure 3: A Terminated Transmission Line [16]

Consider a line terminated by the load Z_L at $z = 0$ as depicted in Figure 3. The characteristic impedance is the ratio of the traveling voltage and current waves.

$$\frac{V_0^+}{I_0^+} = Z_0$$

Substituting the boundary condition $z = 0$, in Equations (17) and (20), we get

$$V(0) = V_0^+ + V_0^- \qquad (31)$$

$$I(0) = \frac{V_0^+}{Z_0} - \frac{V_0^-}{Z_0} \qquad (32)$$

We can define the line impedance $Z(z)$ at any position z on the line as seen in Figure 3:

$$Z(z) = \frac{V(z)}{I(z)} = Z_0 \frac{V_0^+ e^{-\gamma z} + V_0^- e^{\gamma z}}{V_0^+ e^{-\gamma z} - V_0^- e^{\gamma z}} \tag{33}$$

Here, the line impedance is not equal to Z_0 when the line is terminated, i.e., a leftward-traveling reflected wave exists. We can find the line impedance at the load position, i.e., Z_L, by dividing above two equations:

$$\frac{V(z)}{I(z)} \Big|_{z=0} = \frac{V(0)}{I(0)} = Z_L = Z_0 \frac{V_0^+ + V_0^-}{V_0^+ - V_0^-} \tag{34}$$

Now, we define the line impedance in HOL Light as follows:

Definition 6.1. *Line Impedance*
⊢$_{def}$ ∀V1 V2 tlc w z.
  line_impedance V1 V2 tlc w z =
    wave_solution_voltage_phasor V1 V2 tlc w z /
      wave_solution_current_phasor V1 V2 tlc w z

where the HOL Light function `line_impedance` represents the ratio of the total voltage $V(z)$ to the total current $I(z)$ at any position z along the line.

Next, we formally verify that the voltage and current on the transmission line at point $z = 0$ have to abide to the boundary condition imposed by the load.

Theorem 6.1. *Line Impedance at the Load Position ($z = 0$)*
⊢$_{thm}$ ∀V1 V2 R L G C w z.
  let tlc = ((R,L,G,C):trans_line_const) in
  [A1] z = Cx(&0)
    ⇒ line_impedance V1 V2 tlc w z =
      characteristic_impedance tlc w * ((V1 + V2) / (V1 - V2))

The verification of Theorem 6.1 is based on the formalizations of line and characteristic impedances alongside some complex arithmetic reasoning.

We can rearrange Equation (34) as the ratio of the reflected voltage amplitude to the incident voltage amplitude

$$\frac{V_0^-}{V_0^+} = \frac{Z_L - Z_0}{Z_L + Z_0} \tag{35}$$

This ratio of the phasors of the reverse and forward waves at the load position ($z = 0$) is defined as voltage reflection coefficient.

$$\Gamma_L = \frac{V_0^-(0)}{V_0^+(0)} = \frac{V_0^-}{V_0^+} = \frac{Z_L - Z_0}{Z_L + Z_0} \tag{36}$$

Next, we define the voltage reflection coefficient in HOL Light as follows:

Definition 6.2. *Voltage Reflection Coefficient*
$\vdash_{def}$ ∀V1 V2 tlc w z.
voltage_reflection_coefficient V1 V2 tlc w z =
(line_impedance V1 V2 tlc w z - characteristic_impedance tlc w) /
(line_impedance V1 V2 tlc w z + characteristic_impedance tlc w)

Now, we verify that the voltage reflection coefficient is equal to the ratio of reflected voltage to the incident voltage as the following HOL Light theorem:

Theorem 6.2. *Relating Forward-Going Voltage to Reflected Voltage*
$\vdash_{thm}$ ∀V1 V2 R L G C w z.
  let tlc = ((R,L,G,C):trans_line_const) in
  [A1] V1 ≠ V2 ∧ [A2] z = Cx(&0)
  [A3] characteristic_impedance tlc w ≠ Cx(&0)
    ⇒ voltage_reflection_coefficient V1 V2 tlc w z = V2 / V1

Assumption **A1** ensures that voltages are different from each other. Assumption **A2** represents the boundary condition z = 0. Assumption **A3** guarantees that the characteristic impedance is nonzero. The verification of the above theorem is mainly based on Theorem 6.1 along with some complex arithmetic reasoning.

We can also obtain the line impedance at the load ($z = 0$) from the reflection coefficient by rewriting the relationship in Equation (36):

$$Z_L = Z_0 \frac{1 + \Gamma_L}{1 - \Gamma_L} \tag{37}$$

Here, the quantity Γ_L is known as the voltage reflection coefficient. Now, we verify the above relationship as the following HOL Light theorem.

Theorem 6.3. *Final Equation for Line Impedance at the Load Position*
$\vdash_{thm}$ ∀V1 V2 R L G C w z.
  let tlc = ((R,L,G,C):trans_line_const) in
  [A1] V1 ≠ V2 ∧ [A2] z = Cx(&0) ∧
  [A3] characteristic_impedance tlc w ≠ Cx(&0)
  ⇒ line_impedance V1 V2 tlc w z =
     characteristic_impedance tlc w *
       ((Cx(&1) + (voltage_reflection_coefficient V1 V2 tlc w z)) /
         (Cx(&1) - (voltage_reflection_coefficient V1 V2 tlc w z)))

Assumptions A1-A3 are the same as those of Theorem 6.2. The verification of the above theorem is primarily based on Theorems 6.1 and 6.2 alongside some complex arithmetic reasoning.

In the following subsections, we formally analyze short-circuited and open-circuited transmission lines as special cases of a terminated transmission line.

6.1 Short-Circuited Line

When the load end of a transmission line is connected in such a way that it creates a short circuit, it is referred to as a short-circuited transmission line. These lines are extensively used in microwave engineering and Radio-Frequency (RF) systems to ensure a proper impedance matching, which is essential for an efficient power transmission and preserving the integrity of signals. Figure 4 depicts a transmission line of length l that is terminated by a short circuit ensuring a zero load impedance, i.e., $Z_L = 0$.

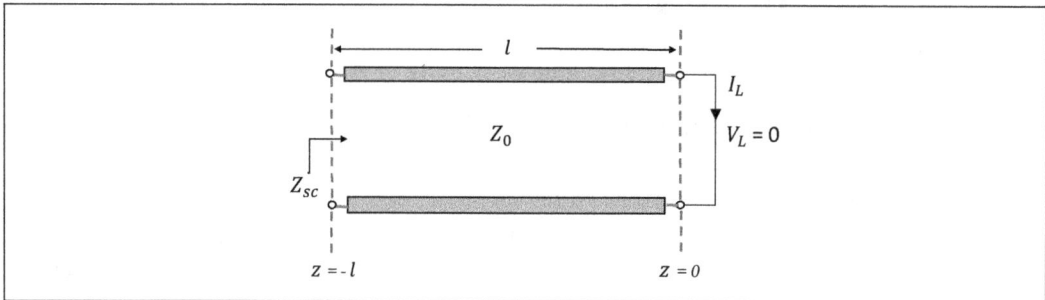

Figure 4: Short-Circuited Line [16]

Moreover, the short-circuited termination forces the load voltage V_L to zero. Therefore, from Equation (17), we have:

$$V_L = V(z)|_{z=0} = 0$$

$$V^+ e^{-j\beta z} + V^- e^{j\beta z}|_{z=0} = 0$$

$$V^+ + V^- = 0$$

This implies

$$V^- = -V^+ \tag{38}$$

228

We employ Equation (20) to find the load current flowing through the short circuit by utilizing Equation (38) as:

$$I_L = I(z)|_{z=0}$$

$$= \frac{1}{Z_0}[V^+ - V^-|_{z=0}$$

$$= \frac{2V^+}{Z_0} \tag{39}$$

Everywhere else on the transmission line, the voltage and current are mathematically expressed as [16]:

$$V(z) = V^+(e^{-j\beta z} - e^{j\beta z}) = -2V^+ j sin(\beta z)$$

$$I(z) = \frac{V^+}{Z_0}(e^{-j\beta z} + e^{j\beta z}) = \frac{2V^+}{Z_0} cos(\beta z)$$

The line impedance observed when looking towards the far end (short-circuited location) on the transmission line is:

$$Z(z) = \frac{V(z)}{I(z)} = Z_0\frac{-2V^+ j sin(\beta z)}{2V^+ cos(\beta z)} = -jZ_0 tan(\beta z)$$

Next, we formally verify the short-circuited line in HOL Light as follows:

Theorem 6.4. *Short-Circuited Line*
$\vdash_{thm}$ ∀V1 V2 R L G C w z.
  let tlc = ((R,L,G,C):trans_line_const) in
  [A1] (V2 = --V1) ∧ [A2] w > &0 ∧ [A3] L > &0 ∧
  [A4] C > &0 ∧ [A5] R = &0 ∧ [A6] G = &0 ∧ [A7] V1 ≠ Cx (&0)
    ⇒ line_impedance V1 V2 tlc w z =
        --ii * characteristic_impedance tlc w *
          ctan (Cx (Im(propagation_constant tlc w)) * z)

Assumptions A1 provides the condition for the short-circuited line. Assumptions A2–A4 guarantee that the angular frequency ω and the parameters L and C cannot be negative or zero, respectively. Assumptions A5–A6 assert that the line parameters R and G are equal to zero, which is an assumptions for a lossless transmission line. Assumption A7 provides that the coefficient V1 is different than zero. The verification of Theorem 6.4 is primarily based on the following lemma:

229

Lemma 6.1. *Lemma for Short-Circuited Line*
$\vdash_{lem}$ ∀V1 V2 R L G C w z.
let tlc = ((R,L,G,C):trans_line_const) in
[A1] (V2 = --V1) ∧ [A2] w > &0 ∧ [A3] L > &0 ∧
[A4] C > &0 ∧ [A5] R = &0 ∧ [A6] G = &0 ∧ [A7] V1 ≠ Cx (&0)
  ⇒ line_impedance V1 V2 tlc w z = characteristic_impedance tlc w *
     ((-Cx(&2) * ii * V1 * csin(Cx (Im(propagation_constant tlc w)) * z)) /
       (Cx(&2) * V1 * ccos (Cx(Im(propagation_constant tlc w)) * z)))

Every assumption in the above lemma is the same as that of Theorem 6.4. The proof of Lemma 6.1 is mainly based on Theorems 1 and 2 provided in Table 4 and properties of the trancendental functions along with some complex arithmetic reasoning.

6.2 Open-Circuited Line

When a transmission line is open at the load end, it is referred to as an open-circuited transmission line. Since the terminal is characterized by an open circuit configuration, the signal or current is unable to propagate beyond the open-circuited point. Open-circuited transmission lines are employed in antenna design to model the behavior of open-ended radiating devices. Figure 5 depicts an open-circuited transmission line with an infinite load impedance, i.e., $Z_L = \infty$.

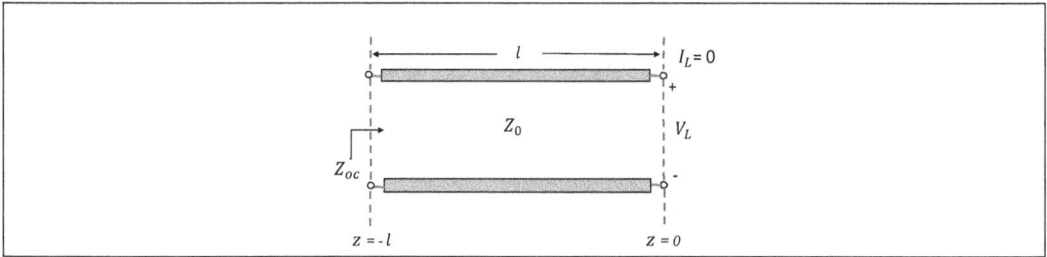

Figure 5: Open-Circuited Line [16]

An open-circuited transmission line forces the load current I_L to be zero. Therefore, by using Equation (20) we have:

$$I_L = I(z)|_{z=0} = 0$$

$$\frac{V^+}{Z_0}e^{-j\beta z} - \frac{V^-}{Z_0}e^{j\beta z}|_{z=0} = 0$$

$$\frac{V^+ + V^-}{Z_0} = 0$$

Thus,

$$V^- = V^+ \qquad (40)$$

Note that the load voltage V_L appearing across the open circuit can be found from Equation (17) using Equation (40):

$$V_L = V(z)|_{z=0}$$

$$= V^+ e^{-j\beta z} + V^- e^{j\beta z}|_{z=0}$$

$$= V^+ + V^- = 2V^+ \qquad (41)$$

Everywhere else on the transmission line, the voltage and current are mathematically expressed as [16]:

$$V(z) = V^+(e^{-j\beta z} + e^{j\beta z}) = 2V^+ cos(\beta z)$$

$$I(z) = \frac{V^+}{Z_0}(e^{-j\beta z} - e^{j\beta z}) = -\frac{2V^+}{Z_0} j sin(\beta z) = \frac{2V^+}{Z_0} e^{-j\pi/2} sin(\beta z)$$

Next, we formally verify the open-circuited line in HOL Light as follows:

Theorem 6.5. *Open-Circuited Line*
$\vdash_{thm} \forall V1\ V2\ R\ L\ G\ C\ w\ z.$
  let tlc = ((R,L,G,C):trans_line_const) in
  [A1] (V2 = V1) $\wedge$ [A2] w > &0 $\wedge$ [A3] L > &0 $\wedge$
  [A4] C > &0 $\wedge$ [A5] R = &0 $\wedge$ [A6] G = &0 $\wedge$ [A7] V1 $\neq$ Cx (&0)
    $\Rightarrow$ line_impedance V1 V2 tlc w z =
        ii * characteristic_impedance tlc w *
          ccot (Cx (Im (propagation_constant tlc w)) * z)

Assumptions **A1** ensures the condition for the open-circuited line. The rest of the assumptions are the same as that of Theorem 6.4. Similar to Theorem 6.4, the proof of the above theorem is mainly based on the following lemma:

Lemma 6.2. *Lemma for Open-Circuited Line*

```
⊢lem ∀V1 V2 R L G C w z.
  let tlc = ((R,L,G,C):trans_line_const) in
  [A1] (V2 = V1) ∧ [A2] w > &0 ∧ [A3] L > &0 ∧
  [A4] C > &0 ∧ [A5] R = &0 ∧ [A6] G = &0 ∧ [A7] V1 ≠ Cx (&0) ∧
  ⇒ line_impedance V1 V2 tlc w z = characteristic_impedance tlc w *
    ((Cx(&2) * V1 * ccos(Cx (Im (propagation_constant tlc w)) * z)) /
    (--Cx(&2) * ii * V1 * csin (Cx (Im (propagation_constant tlc w)) * z)))
```

The proof of the above lemma is mainly based on the formally verified lemmas about the exponential functions alongwith some complex arithmetic reasoning. This completes the formal analysis of the terminated, short-circuited and open-circuited transmission lines. The details about the analysis can be found in the proof script [10].

7 Discussion

The main purpose of this work is the formal development of transmission line theory within the sound core of a higher-order-logic theorem prover to analyze transmission systems. For our constructive formalization, we first formally analyzed the variations of the line voltage and current utilizing the phasor representations of the telegrapher's equations because the phasor approach reduces the time-domain PDEs to ODEs. In the verification of the ODEs, we proved lemmas about the derivatives of the general solutions. One of the main challenges of the presented work was to formally verify the general solutions for the time-domain PDEs. The process began by translating solutions from the phasor domain, where they are articulated as complex-valued functions of frequency, into the time-domain as real-valued functions to establish solutions for PDEs. In the HOL Light proof process, we subsequently faced the requirement to transform the time-domain functions back into complex-valued forms. This was essential because the time-domain PDEs are defined using complex derivatives, and the challenge lays in adeptly employing these complex derivatives during the proof procedure. We also proved the necessary lemmas about the complex differentions of the general solutions with respect to the parameters z and t. In addition, we provided proofs of the attenuation and phase constants for the lossless line and some other theorems regarding exponential functions and complex numbers in order to verify the correctness of the wave solutions for the time-domain PDEs. Once we proved the required theorems and lemmas, the verification of the correctness of the equations just took several lines of proof steps.

For example, the proofs of the general solutions of the wave equations for voltage and current just took 19 and 22 lines, which clearly illustrates the benefit of the

Formalized Theorems	Proof Lines	Page Numbers	Woman Hours
Theorem 4.1	26	212	1
Theorem 4.2	25	212	1
Theorem 5.1	13	214	0.5
Theorem 5.2	25	214	0.5
Table 3. Lemma 1	4	215	0.5
Table 3. Lemma 2	33	215	1
Table 3. Lemma 3	4	215	0.5
Table 3. Lemma 4	57	215	1
Theorem 5.3	28	216	1
Theorem 5.4	49	216	1.5
Table 4. Theorem 1	34	218	3
Table 4. Theorem 2	37	218	4
Table 4. Theorem 3	81	218	3
Table 4. Theorem 4	85	218	7
Table 4. Theorem 5	146	218	2
Table 4. Theorem 6	220	218	3
Theorem 5.5	19	220	0.5
Lemma 5.1	61	220	20
Table 5. Lemma 1	23	221	3
Table 5. Lemma 2	46	221	5
Table 5. Lemma 3	25	221	4
Table 5. Lemma 4	33	221	7
Theorem 5.6	22	222	0.5
Lemma 5.2	82	224	10
Table 6. Lemma 1	54	223	4
Table 6. Lemma 2	32	223	6
Table 6. Lemma 3	59	223	5
Table 6. Lemma 4	38	223	9
Theorem 5.7	71	224	2
Theorem 5.8	107	224	3
Theorem 6.1	18	226	0.5
Theorem 6.2	70	227	3
Theorem 6.3	22	227	0.5
Theorem 6.4	33	229	0.5
Lemma 6.1	63	230	4
Theorem 6.5	48	231	0.5
Lemma 6.2	57	232	5

Table 7: Verification Details for Proven Theorems and Lemmas

formally verified lemmas and theorems. The amount of effort required for verifying each individual theorem in terms of proof lines and the corresponding woman-hours is presented in Table 7. It is noteworthy that the woman-hours needed to complete proofs are dependent on both the number of lines of code and the complexity of the proof. Consequently, there is no direct relation between the number of lines in the proof script and the amount of time required in woman-hours. For example, the verification process for the characteristic impedance of a distortionless line involves a greater number of proof lines compared to the verification of the attenuation constant. However, the woman-hours required for the former are actually less than those needed for the latter. Another difficulty encountered in this formalization pertains to the considerable level of user intervention. However, we developed several tactics that automate certain parts of our proofs resulting in a reduction of the length of proof scripts in many instances (e.g., reducing part of the code by around 240 lines) and make the proofs simpler and more compact. Examples of such tactics are SHORT_TAC and EQ_DIFF_SIMP, which allowed us to simplify complex arithmetics involved in the proof of the time-domain solutions. For instance, EQ_DIFF_SIMP is constructed to efficiently deal with the repetitive patterns in our proof procedure by consolidating them into a single tactic. This proves to be efficient in refining and optimizing our overall approach. The main advantage of the conducted formal proofs of the telegrapher's equations is that all the underlying assumptions can be explicitly written contrary to the case of paper-and-pencil proofs and proof-steps that are mechanically verified using a theorem prover. In addition, the formalization of the transmission line theory provides mathematicians and engineers with the ability to modify and reuse the formal library in HOL Light, in contrast to conventional manual mathematical analysis.

8 Conclusion

This paper advocates the usage of higher-order-logic theorem proving for the formalization of the telegrapher's equations and the verification of its general solutions. In particular, we formalized the telegrapher's equations and their alternate representations, i.e., wave equations in time and phasor domains using HOL Light. Furthermore, we verified the relationship between the telegrapher's equations and the wave equations in the phasor domain. Moreover, we constructed the formal proof for the general solutions of the telegrapher's equations in the phasor domain. Subsequently, we proved the relation between the phasor and the time-domain functions in order to formally verify the general solutions for the time-domain PDEs. Finally, in order to demonstrate the usefulness of our formalization work, we formally analyzed several practical applications, including terminated, short-circuited and open-circuited transmission lines. One of our future plans is to extend this formalization by formalizing the deviations of real circuits from the idealized model, with an aim of applying it to practical applications in real-world scenarios. Another potential area for future investigation involves analyzing the behavior of harmonics in transmission lines using the Fourier transform [18].

References

[1] N. N. Rao, Fundamentals of Electromagnetics for Electrical and Computer Engineering, Pearson Prentice Hall, 2009.

[2] R. J. LeVeque, Finite Difference Methods for Ordinary and Partial Differential Equations: Steady-State and Time-Dependent Problems, SIAM, 2007.

[3] J. Biazar, H. Ebrahimi, Z. Ayati, An approximation to the solution of telegraph equation by variational iteration method, Numerical Methods for Partial Differential Equations 25 (4) (2009) 797–801.

[4] D. Konane, W. Y. S. B. Ouedraogo, T. T. Guingane, A. Zongo, Z. Koalaga, F. Zougmoré, An exact solution of telegraph equations for voltage monitoring of electrical transmission line, Energy and Power Engineering 14 (11) (2022) 669–679.

[5] S. Kühn, General analytic solution of the telegrapher's equations and the resulting consequences for electrically short transmission lines, Journal of Electromagnetic Analysis and Applications 12 (6) (2020) 71–87.

[6] S. Boldo, F. Clément, J.-C. Filliâtre, M. Mayero, G. Melquiond, P. Weis, Trusting computations: a mechanized proof from partial differential equations to actual program, Computers & Mathematics with Applications 68 (3) (2014) 325–352.

[7] W. A. Strauss, Partial Differential Equations: An Introduction., John Wiley & Sons, 2007.

[8] E. Deniz, A. Rashid, O. Hasan, S. Tahar, On the formalization of the heat conduction problem in HOL, in: Intelligent Computer Mathematics, LNCS 13467, Springer, 2022, pp. 21–37.

[9] D. W. Hahn, M. N. Özisik, Heat Conduction, John Wiley & Sons, 2012.

[10] E. Deniz, A. Rashid, Formalization of the Telegrapher's Equations, HOL Light Script, https://hvg.ece.concordia.ca/code/hol-light/pde/te/telegrapher_equations.ml (2023).

[11] J. Harrison, Handbook of Practical Logic and Automated Reasoning, Cambridge University Press, 2009.

[12] D. Leivant, Higher order logic., Handbook of Logic in Artificial Intelligence and Logic Programming (2) (1994) 229–322.

[13] OCaml, https://ocaml.org/ (2023).

[14] L. C. Paulson, ML for the Working Programmer, Cambridge University Press, 1996.

[15] M. N. Sadiku, S. Nelatury, Elements of Electromagnetics, Vol. 428, Oxford University Press, 2001.

[16] U. S. Inan, A. S. Inan, R. K. Said, Engineering Electromagnetics and Waves, Pearson, 2016.

[17] M. B. Steer, Microwave and RF Design: A Systems Approach, SciTech Pub., 2010.

[18] K. Wolf, Integral Transforms in Science and Engineering, Springer Science & Business Media, 2013.

Received July 2023

A Categorical Equivalence for Tense Pseudocomplemented Distributive Lattice

Gustavo Pelaitay

Instituto de Ciencias Básicas, Universidad Nacional de San Juan and CONICET
5400 San Juan, Argentina
gpelaitay@gmail.com

Maia Starobinsky

Facultad de Ciencias Económicas, Universidad de Buenos Aires
Buenos Aires, Argentina
maiastaro@gmail.com

Abstract

In this paper, we introduce the concept of tense operators on pseudocomplemented distributive lattices. Specifically, we utilize the Kalman construction to establish a categorical equivalence between the algebraic category of tense KAN-algebras and a category whose objects are pairs $(\mathbf{A}, S)$, where $\mathbf{A}$ is a tense pseudocomplemented distributive lattice, and S is a tense Boolean filter of $\mathbf{A}$.

1 Introduction

The investigation of tense operators emerged in the 1980s, with notable contributions by Burges (see [4]). Classical tense logic is a logical system that extends bivalent logic by incorporating the tense operators G (indicating that something will always be the case) and H (indicating that something has always been the case) (see [12]). These operators allow us to express statements that hold consistently in the future or have always been true in the past. Tense logic provides a formal framework for reasoning about time-dependent propositions and has applications in various fields, including computer science, artificial intelligence, and philosophy of time.

We extend our sincere gratitude to the editors and anonymous reviewers for their diligent efforts in reviewing and enhancing the quality of this article. Their valuable feedback and constructive suggestions have significantly contributed to the improvement of the manuscript. We appreciate their time, expertise, and commitment to ensuring the excellence of this work.

By incorporating appropriate tense operators, we can expand upon existing logical systems, such as intuitionistic calculus and many-valued logics, to create new tense logics (see [10, 7]). This extension enhances the expressiveness of the logical systems, enabling a more nuanced analysis of the tense dimension in statements. The study of tense logics has led to the development of various variants, each with its own unique features and applications across different fields of study. Two other operators F and P are usually defined via G and H by $F(x) = -G(-x)$ and $P(x) = -H(-x)$, where $-x$ denotes negation of the proposition x. In a classical propositional calculus, which is represented using a Boolean algebra $\mathcal{B} = \langle B, \vee, \wedge, \neg, 0, 1 \rangle$, the axioms for tense operators were established in [12] as follows:

(B1) $G(1) = 1$ and $H(1) = 1$;

(B2) $G(x \wedge y) = G(x) \wedge G(y)$ and $H(x \wedge y) = H(x) \wedge H(y)$;

(B3) $x \leq GP(x)$ and $x \leq HF(x)$.

In order to introduce tense operators in non-classical logics, it is necessary to add additional axioms for G and H to establish their connections with other operations or logical connectives. Tense operators have been extensively investigated by various authors across different classes of algebras (see [1, 3, 5, 8, 9, 10, 14]), and the notion of tense operators on bounded distributive lattices was introduced by Chajda and Paseka in [5]. More precisely, a tense distributive lattice is a structure $\mathcal{A} = \langle A, G, H, F, P \rangle$ where $A = \langle A, \wedge, \vee, 0, 1 \rangle$ is a bounded distributive lattice, and G, H, F, and P are tense operators defined on A. In particular, these operators satisfy:

(T1) $P(x) \leq y$ if and only if $x \leq G(y)$,

(T2) $F(x) \leq y$ if and only if $x \leq H(y)$,

(T3) $G(x) \wedge F(y) \leq F(x \wedge y)$ and $H(x) \wedge P(y) \leq P(x \wedge y)$,

(T4) $G(x \vee y) \leq G(x) \vee F(y)$ and $H(x \vee y) \leq H(x) \vee P(y)$.

Notice that, from the perspective of Universal Algebra, the class of tense distributive lattices constitutes a variety (see [5]).

2 Preliminaries

In this section, we will summarize some definitions and necessary results for what follows. We assume that the reader is familiar with bounded distributive lattices, De Morgan algebras, pseudocomplemented distributive lattices, and Kleene algebras (see [2]).

A pseudocomplemented distributive lattice (or distributive p-algebra) is an algebra $\langle A, \vee, \wedge, ^*, 0, 1 \rangle$ of type $(2, 2, 1, 0, 0)$ such that $\langle A, \vee, \wedge, 0, 1 \rangle$ is a bounded distributive lattice, and for every $a, b \in A$, it holds that $a \wedge b = 0$ if and only if $a \leq b^*$. This means that for every $a \in A$, there is a largest member of A that is disjoint with a, namely a^*. The class of distributive p-algebras is a variety (see [2]). Also, note that in a distributive p-algebra, the conditions $1 = 0^*$ and $0 = 1^*$ necessarily hold.

Recall that if A is a distributive p-algebra, a non-empty subset $S \subseteq A$ is said to be a filter of A if S is an upset, and $x \wedge y \in S$ for all $x, y \in S$.

An element a of A is called a dense element if $a^* = 0$, and the set $D(A)$ of all dense elements of A forms a filter in A.

If A is a distributive p-algebra and R an equivalence relation on A, we adopt the notation $[a]_R$ for the equivalence class of a modulo R, and also A/R for the set of equivalence classes. The definitions of Boolean filter and Boolean congruence on a given distributive p-algebra will be used throughout the paper, so we choose to introduce these definitions and the link between them in the present section (for more details, see [15]).

Definition 2.1. Let A be a distributive p-algebra. We say that a congruence R on A is a Boolean congruence if A/R is a Boolean algebra, or equivalently, if $a \vee a^* \in [1]_R$ for every $a \in A$.

Definition 2.2. Let A be a distributive p-algebra. A filter S of A is called a Boolean filter if $x \vee x^* \in S$ for each x in A.

Since $x \vee x^* \in D(A)$ for all x in A, it is evident that $D(A)$ is a Boolean filter of A. In fact, it is the smallest Boolean filter of A.

The following three lemmas are well-known in the field of distributive p-algebras and are frequently cited, as exemplified in [11, 15].

Lemma 2.1. *Let A be a distributive p-algebra. The following conditions are equivalent:*

1. *A is a Boolean algebra.*

2. *$D(A) = \{1\}$.*

Lemma 2.2. *Let A be a distributive p-algebra and R be a congruence on A. The following conditions are equivalent:*

1. *R is a Boolean congruence.*

2. *$[1]_R$ is a Boolean filter.*

Lemma 2.3. *Let A be a distributive p-algebra. If R is a Boolean congruence, then $[1]_R$ is a Boolean filter. If S is a Boolean filter, then the set*

$$\Theta(S) = \{(a,b) \in A \times A : a \wedge s = b \wedge s \text{ for some } s \in S\}$$

is a Boolean congruence. Moreover, the assignments $R \mapsto [1]_R$ and $S \mapsto \Theta(S)$ define an order isomorphism between the poset of Boolean congruences of A and the poset of Boolean filters of A.

Recall that a Kleene algebra is an algebra $\langle T, \vee, \wedge, \sim, 0, 1\rangle$ of type $(2,2,1,0,0)$ satisfying that $\langle T, \vee, \wedge, 0, 1\rangle$ is a bounded distributive lattice and $\sim$ is an involution (i.e., $\sim\sim x = x$ for every $x \in T$) such that

1. $\sim (x \vee y) = \sim x \wedge \sim y$ and

2. $x \wedge \sim x \leq y \vee \sim y$.

hold for every $x, y \in T$.

In [11], the authors extend Kleene algebras with a unary operation $\neg$, referred to as intuitionistic negation, and define the variety of Kleene algebras with intuitionistic negation, abbreviated as KAN-algebras. More precisely, a KAN-algebra is an algebra $\langle T, \wedge, \vee, \sim, \neg, 0, 1\rangle$ of type $(2,2,1,1,0,0)$ such that $\langle T, \wedge, \vee, \sim, 0, 1\rangle$ is a Kleene algebra and the following conditions are satisfied for every $x, y \in T$:

(N1) $\neg(x \wedge \neg(x \wedge y)) = \neg(x \wedge \neg y)$,

(N2) $\neg(x \vee y) = \neg x \wedge \neg y$,

(N3) $x \wedge \sim x = x \wedge \neg x$,

(N4) $\sim x \leq \neg x$,

(N5) $\neg(x \wedge y) = \neg((\sim \neg x) \wedge y)$.

If $\langle T, \vee, \wedge, \sim, \neg, 0, 1\rangle$ is a KAN-algebra, an application of (N3) yields $\neg 1 = 1 \wedge \neg 1 = 1 \wedge \sim 1 = 1 \wedge 0 = 0$. Taking $x = 0$ in (N4) we obtain that $\neg 0 = 1$. In addition, if $x \leq y$, then $\neg y \leq \neg x$ by (N2).

3 Tense operators on distributive p-algebras

In this section, we will define the variety of tense pseudocomplemented distributive lattices and prove some basic properties. Additionally, we will introduce the tense version of Boolean filter and Boolean congruence in the subsequent discussion.

Definition 3.1. An algebra $\mathbf{A} = (A, G, H, F, P)$ is a tense pseudocomplemented distributive lattice, or tense p-algebra, if $\langle A, \vee, \wedge,^*, 0, 1 \rangle$ is a distributive p-algebra, and G, H, F, P are unary operations on A that satisfy the following conditions:

(T1) $P(x) \leq y$ if and only if $x \leq G(y)$,

(T2) $F(x) \leq y$ if and only if $x \leq H(y)$,

(T3) $G(x) \wedge F(y) \leq F(x \wedge y)$ and $H(x) \wedge P(y) \leq P(x \wedge y)$,

(T4) $G(x \vee y) \leq G(x) \vee F(y)$ and $H(x \vee y) \leq H(x) \vee P(y)$,

(T5) $F(x)^* \leq G(x^*)$ and $P(x)^* \leq H(x^*)$,

(T6) $G(x)^* \leq F(x^*)$ and $H(x)^* \leq P(x^*)$.

***Example* 3.1.** Given a distributive p-algebra A, there are two extreme examples of tense operators:

(1) Define $G, H, F,$ y P as the identity function id_A.

(2) Define G and H as the constant function 1_A (i.e., $G(x) = 1 = H(x)$ for all $x \in A$), and F and P as the constant function 0_A (i.e., $F(x) = 0 = P(x)$ for all $x \in A$).

Remark 3.1. Let $\mathbf{A} = (A, G, H, F, P)$ be a tense p-algebra. Then, according to properties (T1) to (T4), we can conclude that the reduct $\langle A, \vee, \wedge, G, H, F, P \rangle$ forms a tense distributive lattice (see [5, 13]).

We will list several fundamental properties that hold in tense p-algebras and provide proofs for some of them.

Proposition 3.1. *Let* (A, G, H, F, P) *be a tense p-algebra. Then*

(T7) $G(1) = 1$ *and* $H(1) = 1$,

(T8) $G(x \wedge y) = G(x) \wedge G(y)$ *and* $H(x \wedge y) = H(x) \wedge H(y)$,

(T9) $x \leq GP(x)$ and $x \leq HF(x)$,

(T10) $F(0) = 0$ and $P(0) = 0$,

(T11) $F(x \vee y) = F(x) \vee F(y)$ and $P(x \vee y) = P(x) \vee P(y)$,

(T12) $FH(x) \leq x$ and $PG(x) \leq x$,

(T13) $x \leq y$ implies $G(x) \leq G(y)$ and $H(x) \leq H(y)$,

(T14) $x \leq y$ implies $F(x) \leq F(y)$ and $P(x) \leq P(y)$,

(T15) $x \wedge F(y) \leq F(P(x) \wedge y)$ and $x \wedge P(y) \leq P(F(x) \wedge y)$,

(T16) $F(x) \wedge y = 0$ if and only if $x \wedge P(y) = 0$,

(T17) $G(x \vee H(y)) \leq G(x) \vee y$ and $H(x \vee G(y)) \leq H(x) \vee y$,

(T18) $x \vee H(y) = 1$ if and only if $G(x) \vee y = 1$,

(T19) $G(x^*) \leq F(x)^*$ and $H(x^*) \leq P(x)^*$,

(T20) $F(x^*) \leq G(x)^*$ and $P(x^*) \leq H(x)^*$.

Proof. Note that (T7) to (T12) follow from (T1) and (T2). Axioms (T13) and (T14) are consequences of axioms (T8) and (T11), respectively. Next, let's prove (T15). From (T9), we have $x \wedge F(y) \leq GP(x) \wedge F(y)$. Using this statement and (T3), we obtain $x \wedge F(y) \leq F(P(x) \wedge y)$. The reverse inequality can be proven similarly. Now, let's verify (T16). Suppose $F(x) \wedge y = 0$. Using (T10) and (T15), we get $x \wedge P(y) \leq P(F(x) \wedge y) = P(0) = 0$. Hence, $x \wedge P(y) = 0$. Similarly, we can prove the reverse direction. Moreover, axioms (T17) and (T18) can be proven using a similar technique as in the proof of (T15) and (T16), respectively. Finally, let's prove (T19) and (T20). Using (T3) and (T10), we have $G(x^*) \wedge F(x) \leq F(x^* \wedge x) = F(0) = 0$. Thus, $G(x^*) \leq F(x)^*$. Similarly, $H(x^*) \leq P(x)^*$. Additionally, (T20) can be proven using a similar technique. $\qquad\square$

Remark 3.2. If $\mathbf{A} = (A, G, H, F, P)$ is a tense p-algebra, and A is a Boolean algebra, it is easy to see that (A, G, H) is a tense Boolean algebra.

Definition 3.2. Let $\mathbf{A} = (A, G, H, F, P)$ be a tense p-algebra. A filter S of A is called a *tense filter* if it satisfies the following condition:

(tf) $x \in S$ implies $G(x) \in S$ and $H(x) \in S$.

Definition 3.3. Let $\mathbf{A} = (A, G, H, F, P)$ be a tense p-algebra. A tense filter $S \subseteq A$ is called a *tense Boolean filter* if it contains all dense elements, i.e., $D(A) \subseteq S$.

Example 3.2. Let $\mathbf{A} = (A, G, H, F, P)$ be a tense p-algebra. The set $D(A)$ is a tense Boolean filter. It is evident that $D(A)$ forms a Boolean filter. Let's prove that $D(A)$ is closed under both G and H. Suppose $x \in D(A)$. From this assertion, applying axioms (T6), (T20), and (T10), we have $G(x)^* = F(x^*) = F(0) = 0$. Consequently, it follows that $G(x) \in D(A)$ Similarly, we can verify that $H(x) \in D(A)$.

Definition 3.4. Let $\mathbf{A} = (A, G, H, F, P)$ be a tense p-algebra. A *tense congruence* on $\mathbf{A}$ is a p-congruence θ which is compatible with every tense operators, i.e. if $(x, y) \in \theta$, then $(T(x), T(y)) \in \theta$, for every $T \in \{G, H, F, P\}$.

Definition 3.5. Let $\mathbf{A} = (A, G, H, F, P)$ be a tense p-algebra. A tense congruence θ is a *tense Boolean congruence* of $\mathbf{A}$ if the quotient algebra $\mathbf{A}/\theta = (A/\theta, G, H)$ is a tense Boolean algebra.

Remark 3.3. Let $\mathbf{A} = (A, G, H, F, P)$ be a tense p-algebra. The set of all tense Boolean congruences forms a lattice.

Lemma 3.1. *Let $\mathbf{A} = (A, G, H, F, P)$ be a tense p-algebra. If θ is a tense Boolean congruence, then $[1]_\theta$ is a tense Boolean filter of $\mathbf{A}$.*

Proof. It is known that $[1]_\theta$ is a Boolean filter (see [11, Lemma 1.2]). Let $x \in [1]_\theta$. Then, $(x, 1) \in \theta$. Since θ is a tense congruence, we have $(G(x), G(1)) \in \theta$. Applying property (T8), we conclude that $G(1) = 1$. Therefore, $G(x) \in [1]_\theta$. Similarly, we can deduce that $H(x) \in [1]_\theta$ using a similar approach. Therefore, $[1]_\theta$ is a tense Boolean filter. $\square$

From the established results in [11] and Lemma 3.1, the following result is obtained.

Lemma 3.2. *Let $\mathbf{A} = (A, G, H, F, P)$ be a tense p-algebra and S a tense Boolean filter. Then, the set*

$$\Theta(S) = \{(a, b) \in A \times A : a \wedge s = b \wedge s \text{ for some } s \in S\}$$

is a tense Boolean congruence. Moreover, the assignments $\theta \mapsto [1]_\theta$ and $S \mapsto \Theta(S)$ define an order isomorphism between the poset of tense Boolean congruences of $\mathbf{A}$ and the poset of tense Boolean filters of $\mathbf{A}$.

Remark 3.4. Upon examining the assignments from the previous lemma, it can be proven that a correspondence exists between the set of all tense filters and the set of all tense congruences of a tense p-algebra $\mathbf{A}$.

4 Tense operators on KAN-algebras

In this section we will introduce the notion of tense operators on the variety of KAN-algebras.

Let $\langle T, \vee, \wedge, \sim, \neg, 0, 1 \rangle$ be a KAN-algebra, and let G and H be two unary operators on T. We define the operators $P(x) := \sim H(\sim x)$ and $F(x) := \sim G(\sim x)$.

Definition 4.1. An algebra $\mathbf{T} = (T, G, H)$ is a tense KAN-algebra if $\langle T, \vee, \wedge, \sim, \neg, 0, 1 \rangle$ is a KAN-algebra, and G and H are unary operations on T that satisfy the following conditions:

(t1) $G(1) = 1$ and $H(1) = 1$,

(t2) $G(x \wedge y) = G(x) \wedge G(y)$ and $H(x \wedge y) = H(x) \wedge H(y)$,

(t3) $x \leq GP(x)$ and $x \leq HF(x)$,

(t4) $G(x \vee y) \leq G(x) \vee F(y)$ and $H(x \vee y) \leq H(x) \vee P(y)$,

(t5) $G(\neg x) = \neg F(x)$ and $P(\neg x) = \neg H(x)$,

(t6) $\neg G(x) = F(\neg x)$ and $H(\neg x) = \neg P(x)$.

***Example* 4.1.** Let $\mathbf{B} = (B, G, H)$ be a tense Boolean algebra, and let the unary operation $\sim$ be defined as $\sim x := \neg x$. According to Example 2.3 in [11], it is stated that $(B, \wedge, \vee, \sim, \neg, 0, 1)$ is a KAN-algebra. By checking that G and H satisfy the axioms (t1) to (t6), we can conclude that $\mathbf{B}$, with this additional operation $\sim$, is a tense KAN-algebra.

The following proposition contains some properties of tense KAN-algebras that will be useful throughout the paper. The proof is analogous to that of Lemma 3.1, so we omit it.

Proposition 4.1. *Let* $\mathbf{T} = (T, G, H)$ *be a tense KAN-algebra. Then,*

(t7) $F(0) = 0$ *and* $P(0) = 0$,

(t8) $F(x \vee y) = F(x) \vee F(y)$ *and* $P(x \vee y) = P(x) \vee P(y)$,

(t9) $PG(x) \leq x$ *and* $FH(x) \leq x$,

(t10) $x \leq y$ *implies* $G(x) \leq G(y)$ *and* $H(x) \leq H(y)$,

(t11) $x \leq y$ *implies* $F(x) \leq F(y)$ *and* $P(x) \leq P(y)$,

(t12) $G(x) \wedge F(y) \leq F(x \wedge y)$ *and* $H(x) \wedge P(y) \leq P(x \wedge y)$, (t13) $x \wedge F(y) \leq F(P(x) \wedge y)$ and $x \wedge P(y) \leq P(F(x) \wedge y)$,

(t14) $F(x) \wedge y = 0$ if and only if $x \wedge P(y) = 0$,

(t15) $G(x \vee H(y)) \leq G(x) \vee y$ and $H(x \vee G(y)) \leq H(x) \vee y$,

(t16) $x \vee H(y) = 1$ if and only if $G(x) \vee y = 1$.

5 Kalman's Construction

In this section, we prove some results that establish the connection between tense p-algebras and tense KAN-algebras.

Let $\mathbf{A} = (A, G, H, F, P)$ be a tense p-algebra and let us consider

$$K(A) := \{(a,b) \in A \times A : a \wedge b = 0\}.$$

As established in the well-known [11, Lemma 2.4], by defining:

$$
\begin{aligned}
(a,b) \vee (x,y) &:= (a \vee x, b \wedge y), \\
(a,b) \wedge (x,y) &:= (a \wedge x, b \vee y), \\
\neg(a,b) &:= (a^*, a), \\
\sim(a,b) &= (b,a), \\
0 &= (0,1), \\
1 &= (1,0),
\end{aligned}
$$

we get that the algebra $\mathbb{K}(A) = \langle K(A), \vee, \wedge, \sim, \neg, 0, 1 \rangle$ is a KAN-algebra.

Now, let us define the following unary operators on $K(A)$:

$$
\begin{aligned}
G_K((a,b)) &:= (G(a), F(b)), \\
H_K((a,b)) &:= (H(a), P(b)), \\
F_K((a,b)) &:= (F(a), G(b)), \\
P_K((a,b)) &:= (P(a), H(b)).
\end{aligned}
$$

Lemma 5.1. Let $\mathbf{A} = (A, G, H, F, P)$ be a tense p-algebra and let $(a,b) \in K(A)$. Then, the following hold:

(a) $G_K(a,b) \in K(A)$ and $H_K(a,b) \in K(A)$,

(b) $F_K(a, b) =\sim G_K(\sim (a, b))$ and $P_K(a, b) =\sim H_K(\sim (a, b))$,

(c) $F_K(a, b) \in K(A)$ and $P_K(a, b) \in K(A)$.

Proof. We will focus on proving property (a), leaving the remaining properties for the reader to verify. Let $(a, b) \in K(A)$. Hence, $a \wedge b = 0$. Then, from (T3) and (T10), $G(a) \wedge F(b) \leq F(a \wedge b) = F(0) = 0$. Therefore, $(G(a), F(b)) \in K(A)$. In a similar way, we can prove $H_K(a, b) \in K(A)$. $\qquad\square$

Lemma 5.2. *Let* $\mathbf{A} = (A, G, H, F, P)$ *be a tense p-algebra. Then, the structure* $K(\mathbf{A}) = (\mathbb{K}(A), G_K, H_K)$ *is a tense KAN-algebra. Furthermore, for* $\mathbf{B} = (B, G, H, F, P)$ *as a tense p-algebra and a morphism* $f : \mathbf{A} \longrightarrow \mathbf{B}$, *the map* $K(f) : K(\mathbf{A}) \longrightarrow K(\mathbf{B})$, *defined by* $K(f)(a, b) = (f(a), f(b))$, *is a functor from the category of tense p-algebras to the category of tense KAN-algebras.*

Proof. Based on [11, Lemma 2.4], we are aware that $\mathbb{K}(A)$ is a KAN-algebra, and from Lemma 5.1 we know that G_K and H_K are well-defined. Therefore, our focus will be on proving that $K(\mathbf{A})$ satisfies axioms (t1) to (t6). Due to the symmetry of tense operators G and H, we will only prove the axioms for the operator G.

Let (a, b) and (x, y) be elements of $K(\mathbf{A})$.

(t1): $G_K(1, 0) = (G(1), F(0)) = (1, 0)$.

(t2): $G_K((a, b) \wedge (x, y)) = G_K(a \wedge x, b \vee y) = (G(a \wedge x), F(b \vee y))$. Using (T8) and (T11), we have $(G(a\wedge x), F(b\vee y)) = (G(a)\wedge G(x), F(b)\vee F(y)) = (G(a), F(b))\wedge (G(x), F(y))$. Therefore, $G_K((a, b) \wedge (x, y)) = G_K(a, b) \wedge G_K(x, y)$.

(t3): $G_K(P_K(a, b)) = G_K(P(a), H(b)) = (G(P(a)), F(H(b)))$. Using (T9) and (T12), we have $(a, b) \leq (G(P(a)), F(H(b)))$, hence $(a, b) \leq G_K P_K(a, b)$.

(t4): $G_K((a, b) \vee (x, y)) = G_K(a \vee x, b \wedge y) = (G(a \vee x), F(b \wedge y))$. Using (T3) and (T4), we have $(G(a \vee x), F(b \wedge y)) \leq (G(a) \vee F(x), G(y) \wedge F(b))$, and $(G(a)\vee F(x), G(y)\wedge F(b)) = (G(a), F(b))\vee (F(x), G(y)) = G_K(a, b)\vee F_K(x, y)$. Therefore, $G_K((a, b) \vee (x, y)) \leq G_K(a, b) \vee F_K(x, y)$.

(t5): $\neg F_K(a, b) = \neg(F(a), G(b)) = (F(a)^*, F(a))$. From (T5) and (T19), we have $F(a)^* = G(a^*)$, thus $\neg F_K(a, b) = (G(a^*), F(a)) = G_K(a^*, a) = G_K(\neg(a, b))$.

(t6): $F_K(\neg(a, b)) = F_K(a^*, a) = (F(a^*), G(a))$. Using (T6) and (T20), we have $F(a^*) = G(a)^*$, therefore $F_K(\neg(a, b)) = (G(a)^*, G(a)) = \neg G_K(a, b)$.

Now, according to [11, Lemma 2.4], we know that K is a functor from the category of distributive p-algebras to the category of KAN-algebras. We will prove that K preserves the tense operator G: $K(f)(G_K(a,b)) = K(f)(G(a), F(b)) = (f(G(a)), f(F(b))) = K(f)(G(a), F(b)) = K(f)(G_K(a,b))$. Similarly, we can observe that K preserves H, F, and P. □

Let $(T, \wedge, \vee, \sim, \neg, 0, 1)$ be a KAN-algebra, and let $\theta \subseteq T^2$ be defined as

$$(x, y) \in \theta \Longleftrightarrow \neg x = \neg y \tag{1}$$

The relation θ is an equivalence relation that will play a crucial role in establishing a categorical equivalence for the class of tense KAN-algebras.

Recall that $[x]_\theta$ denotes the set $\{y \in T : (x, y) \in \theta\}$, and the set $\{[x]_\theta : x \in T\}$ is denoted by T/θ.

Lemma 5.3. *Let* $\mathbf{T} = (T, G, H)$ *be a tense KAN-algebra, and let* $\theta \subseteq T^2$ *be defined as specified in 1. Then, the equivalence relation* θ *is compatible with the operations* $\wedge, \vee, \neg$, *as well as the tense operators* G *and* H.

Proof. From [11, Lemma 2.7], we know that θ is compatible with $\wedge, \vee$, and $\neg$. We will now prove that θ is also compatible with the tense operators G and H. Let $(x, y) \in \theta$. We have $\neg x = \neg y$, which implies $F(\neg x) = F(\neg y)$. By applying property (t6), we have $\neg G(x) = \neg G(y)$, and therefore $(G(x), G(y)) \in \theta$. Similarly, we can show that $(H(x), H(y)) \in \theta$. This confirms that θ is compatible with the tense operators. □

Let $\mathbf{T} = (T, G, H)$ be a tense KAN-algebra. By applying Lemma 5.3 and [11, Lemma 1.8], we deduce that $(T/\theta, \wedge, \vee, \neg, [0]_\theta, [1]_\theta)$ forms a distributive p-algebra, and the order $\leq$ in T/θ can be characterized as $[x]_\theta \leq [y]_\theta$ if and only if $\neg y \leq \neg x$.

Lemma 5.4. *Let* $\mathbf{T} = (T, G, H)$ *be a tense KAN-algebra, and consider the relation* θ *defined in 1. By defining* $G_\theta([x]_\theta) = [G(x)]_\theta$, $H_\theta([x]_\theta) = [H(x)]_\theta$, $F_\theta([x]_\theta) = [F(x)]_\theta$, *and* $P_\theta([x]_\theta) = [P(x)]_\theta$, *we have that* $(T/\theta, G_\theta, H_\theta, F_\theta, P_\theta)$ *forms a tense p-algebra.*

Proof.

(T1): Let's assume $P_\theta([x]_\theta) \leq [y]_\theta$. Due to the characterization of the order in T/θ, it follows that $\neg y \leq \neg P(x)$. Using property (t11), we have $F(\neg y) \leq F(\neg P(x))$, and by applying properties (t6) and (t3), we obtain $\neg G(y) \leq \neg x$. Consequently, $[x]_\theta \leq G_\theta([y]_\theta)$. Now, let's assume $[x]_\theta \leq G_\theta([y]_\theta)$, which implies $\neg G(y) \leq \neg x$. From property (t10), we have $H(\neg G(y)) \leq H(\neg x)$, and by using properties (t6) and (t11), we obtain $\neg y \leq \neg P(x)$. Hence, $P_\theta([x]_\theta) \leq [y]_\theta$.

(T2): The proof is analogous to (T1).

(T3): It is immediate from property (t12) and the fact that $x \leq y$ implies $\neg y \leq \neg x$.

(T4): It is immediate from property (t4) and the fact that $x \leq y$ implies $\neg y \leq \neg x$.

(T5): From property (t5), we have $\neg G(\neg x) = \neg\neg F(x)$, which implies $\neg F_\theta(x) \leq G_\theta(\neg x)$. Similarly, we have $\neg P_\theta(x) \leq H_\theta(\neg x)$.

(T6): The proof is similar to the proof of (T5).

$\square$

Lemma 5.5. *Let* $\mathbf{A} = (A, G, H, F, P)$ *be a tense p-algebra. Then the mapping* $g : K(\mathbf{A})/\theta \longrightarrow \mathbf{A}$ *defined as follows:*

$$g([(a, b)]_\theta) = a,$$

is an isomorphism of tense p-algebras.

Proof. We will only prove that g preserves the tense operators. We have that $g(G_\theta([(a, b)]_\theta)) = g([G_K(a, b)]_\theta) = g([(G(a), F(b))]_\theta) = G(a)$, and $G([g(a, b)]_\theta) = G(a)$. Similarly, we can prove that g preserves H_θ, F_θ, and P_θ. Hence, g is an isomorphism of tense p-algebras. $\square$

Lemma 5.6. *Let* $\mathbf{T} = (T, G, H)$ *be a tense KAN-algebra. Then the mapping* $\rho : \mathbf{T} \longrightarrow K(\mathbf{T}/\theta)$ *defined as* $\rho(x) = ([x]_\theta, ([\sim x]_\theta)$ *is an injective morphism of tense KAN-algebras.*

Proof. We will show that ρ preserves the tense operators G and H. We have that $\rho(G(x)) = ([G(x)]_\theta, [\sim G(x)]_\theta)$, and

$$G_K(\rho(x)) = G_K([x]_\theta, [\sim x]_\theta) = (G_\theta([x]_\theta), F(([\sim x])_\theta) = ([G(x)]_\theta, [F(\sim x)]_\theta).$$

Since $F(\sim x) = \sim G(x)$, we can conclude that ρ preserves G. Similarly, we can prove that ρ preserves H. Therefore, ρ is an injective morphism of tense KAN-algebras. $\square$

6 A categorical equivalence for tense KAN-algebras

In this section, we will prove that tense p-algebras and tense Boolean filters provide a characterization of tense KAN-algebras.

If A is a p-algebra and S a Boolean filter of A, we can define the set

$$K(A, S) := \{(a, b) \in A \times A : a \wedge b = 0 \text{ and } a \vee b \in S\}.$$

From [11, Theorem 2.11], we know that for two p-algebras A and A', and two Boolean filters S and S', there exists a well-defined function $K(f) : K(A, S) \longrightarrow K(A', S')$ given by $K(f)(a, b) = (f(a), f(b))$. Furthermore, it is known that the set $K(A, S)$ is the universe of a subalgebra of $\mathbb{K}(A)$, making it a KAN-algebra.

Proposition 6.1. *If* $\mathbf{A} = (A, G, H, F, P)$ *is a tense p-algebra and S is a tense Boolean filter of A, then the set $K(\mathbf{A}, \mathbf{S}) := \{(a, b) \in A \times A : a \wedge b = 0 \text{ and } a \vee b \in S\}$ is a tense KAN-algebra.*

Proof. We know that $K(A, S)$ is a subalgebra of $\mathbb{K}(A)$. Therefore, we only need to prove that $K(\mathbf{A}, \mathbf{S})$ is closed under the tense operators G_K and H_K. Let $(a, b) \in K(\mathbf{A}, \mathbf{S})$. We have $G_K(a, b) = (G(a), F(b))$. By using (t12) and (t7), we have $G(a) \wedge F(b) \leq F(a \wedge b) = 0$. Additionally, using (t4) and the fact that S is a tense filter, we have that $G(a) \vee F(b) \in S$. Therefore, $G_K(a, b) \in K(\mathbf{A}, \mathbf{S})$. The proof for H_K follows a similar argument. $\square$

Proposition 6.2. *If* $\mathbf{A}$ *and* $\mathbf{A}'$ *are two tense p-algebras, and S and S' are two tense Boolean filters of $\mathbf{A}$ and $\mathbf{A}'$ respectively, let $f : \mathbf{A} \longrightarrow \mathbf{A}'$ be a morphism of tense p-algebras such that $f(S) \subseteq S'$. We can define the morphism $K(f) : K(\mathbf{A}, \mathbf{S}) \longrightarrow K(\mathbf{A}', \mathbf{S}')$ of tense KAN-algebras as $K(f)(a, b) = (f(a), f(b))$.*

Proof. From [11, Theorem 2.11], we know that $K(f)$ is a morphism of KAN-algebras. By Proposition 6.1, we have that $K(\mathbf{A}, \mathbf{S})$ is a tense KAN-algebra, so we only need to prove that $K(f)$ preserves the tense operators G_K and H_K. For G_K, we have $K(f)(G_K(a, b)) = K(f)(G(a), F(b)) = (f(G(a)), f(F(b)))$. Since f is a morphism of tense p-algebras, it preserves G and F. Therefore, $(f(G(a)), f(F(b))) = (G(f(a)), F(f(b))) = G_K(f(a), f(b)) = G_K(K(f)(a, b))$. Hence, $K(f)$ preserves G_K. The proof for H_K follows a similar argument. $\square$

The proof of the following theorem can be obtained by combining the results from Lemma 5.2, Lemma 5.5, and [11, Theorem 2.11].

Theorem 6.1. *If* $\mathbf{A}$ *is a tense p-algebra and S is a tense Boolean filter of $\mathbf{A}$, then the quotient algebra $K(\mathbf{A}, \mathbf{S})/\theta$ is isomorphic to $\mathbf{A}$. Furthermore, if $\mathbf{A}'$ is a tense p-algebra and S' is a tense Boolean filter of $\mathbf{A}'$, and $f : \mathbf{A} \longrightarrow \mathbf{A}'$ is a morphism of tense p-algebras such that $f(S) \subseteq S'$, then $K(f)$ is a morphism of tense KAN-algebras.*

Lemma 6.1. *Let* $\mathbf{T} = (T, G, H)$ *be a tense KAN-algebra . Then the positive part $\mathbf{T}^+ := \{x \in T : \neg \sim x = 1\}$ is a tense filter of $\mathbf{T}$ that includes all elements $x \in \mathbf{T}$ satisfying $\neg\neg x = 1$. Consequently, $\mathbf{T}^+/\theta$ is a tense Boolean filter of $\mathbf{T}/\theta$.*

Proof. From [11, Lemma 2.14], we only need to prove that $\mathbf{T}^+$ is closed under G and H. Suppose $x \in \mathbf{T}^+$. Then, we have that $\neg \sim x = 1$. Therefore, $G(\neg \sim x) = 1$. By using the definition of F and property (t5), we have $\neg F(\sim x) = \neg \sim G(x) = G(\neg \sim x) = 1$. Consequently, $\neg \sim G(x) = 1$, which implies $G(x) \in T^+$. The proof for H follows a similar reasoning. Hence, $\mathbf{T}^+$ is a tense filter of $\mathbf{T}$, and consequently, $\mathbf{T}^+/\theta$ forms a tense filter of $\mathbf{T}/\theta$. $\qquad\square$

Theorem 6.2. *Let* $\mathbf{T} = (T, G, H)$ *be a tense KAN-algebra. Then* $\mathbf{T}$ *is isomorphic to* $K(\mathbf{T}/\theta; \mathbf{T}^+/\theta)$. *Furthermore, if* $\mathbf{T}'$ *is also a tense KAN-algebra and* $f : \mathbf{T} \longrightarrow \mathbf{T}'$ *is a morphism between tense KAN-algebras, then the mapping* $f_\theta : \mathbf{T}/\theta \longrightarrow \mathbf{T}'/\theta$ *defined as* $f_\theta([x]_\theta) = [f(x)]_\theta$ *is a morphism of tense p-algebras. It is worth noting that* $f_\theta(\mathbf{T}^+/\theta) \subseteq (\mathbf{T}')^+/\theta$ *holds.*

Proof. We know from [11, Theorem] that $\rho : T \longrightarrow K(T/\theta; T^+/\theta)$ is an isomorphism of KAN-algebras, and Lemma 5.6 establishes that ρ is a tense morphism. Therefore, ρ is an isomorphism for tense KAN-algebras. Furthermore, according to [11, Theorem 2.15], we only need to proof that f_θ preserves the tense operators. Let's consider $f_\theta(G_\theta([x]_\theta)) = f_\theta([G(x)]_\theta) = [f(G(x))]_\theta$. Since f is a morphism of tense p-algebras, we have $[f(G(x))]_\theta = [G(f(x))]_\theta = G_\theta([f(x)]_\theta) = G_\theta(f_\theta([x]_\theta))$. Hence, we conclude that $f_\theta(G_\theta([x]_\theta)) = G_\theta(f_\theta([x]_\theta))$. Similar reasoning can be applied to prove the preservation of tense operators for H_θ, F_θ, and P_θ. $\qquad\square$

We denote by **tPDL** the category whose objects are pairs $(\mathbf{A}, S)$, where $\mathbf{A}$ is a tense p-algebra and S is a tense Boolean filter of A, and whose arrows $f : (\mathbf{A}, S) \longrightarrow (\mathbf{A}', S')$ are morphisms $f : \mathbf{A} \longrightarrow \mathbf{A}'$ such that $f(S) \subseteq S'$.

Based on the previous results, we can conclude that if $\mathbf{T} = (T, G, H)$ is a tense KAN-algebra, then $K(\mathbf{T}/\theta, \mathbf{T}^+/\theta) \in$ **tPDL**. Moreover, when $f : \mathbf{T} \longrightarrow \mathbf{T}'$ is a morphism between tense KAN-algebras, it follows that f_θ is a morphism in **tPDL**. Consequently, we can observe that the aforementioned assignments establish a functor from the algebraic category of tense KAN-algebras to the category **tPDL**.

The proof of the following theorem is a direct consequence of Theorem 6.1 and Theorem 6.2.

Theorem 6.3. *The functor K establishes a categorical equivalence between the category of tense KAN-algebras and the category* **tPDL**.

References

[1] Almiñana, Federico G.; Pelaitay, Gustavo; Zuluaga, William. *On Heyting Algebras with Negative Tense Operators.* Studia Logica 111 (2023), no. 6, 1015–1036.

[2] Balbes, R., Dwinger P., Distributive Lattices. University of Missouri Press (1974).

[3] Bakhshi, M. Tense operators on non-commutative residuated lattices. *Soft Comput* 21, 4257–4268 (2017).

[4] Burgess, John P. Basic tense logic. Handbook of philosophical logic, Vol. II, 89–133, Synthese Lib., 165, Reidel, Dordrecht, 1984.

[5] Chajda I., Paseka J.: Algebraic Approach to Tense Operators, Research and Exposition in Mathematics Vol. 35, Heldermann Verlag (Germany), 2015, ISBN 978-3-88538-235-5.

[6] Diaconescu, Denisa; Georgescu, George. Tense operators on MV-algebras and Łukasiewicz-Moisil algebras. *Fund. Inform.* 81 (2007), no. 4, 379–408.

[7] Ewald, W. B. Intuitionistic tense and modal logic. *J. Symbolic Logic* 51 (1986), no. 1, 166–179.

[8] Figallo, A.V., Pascual, I. & Pelaitay, G. A topological duality for tense θ-valued Łukasiewicz–Moisil algebras. *Soft Comput* 23, 3979–3997 (2019).

[9] Gallardo, Carlos; Pelaitay, Gustavo; Gallardo, Cecilia Segura. T-rough symmetric Heyting algebras with tense operators. *Fuzzy Sets and Systems* 466 (2023), Paper No. 108455, 13 pp.

[10] Ghorbani, S. Tense operators on frameable equality algebras. *Soft Comput* 26, 203–213 (2022).

[11] Gomez, Conrado; Marcos, Miguel Andrés; San Martín, Hernán Javier. On the relation of negations in Nelson algebras. *Rep. Math. Logic* No. 56 (2021), 15–56.

[12] Kowalski T., Varieties of tense algebras, *Reports on Mathematical Logic*, vol. 32 (1998), pp. 53–95.

[13] Menni, M. and Smith, C., Modes of adjointness. *J. Philos. Logic* 43 (2014), no. 2-3, 365–391.

[14] Paad, Akbar. Tense operators on BL-algebras and their applications. *Bull. Sect. Logic Univ. Łódź* 50 (2021), no. 3, 299–324.

[15] Rao, M.S. and Shum, K., Boolean filters of distributive lattices, *International Journal of Mathematics and Soft Computing* 3:3 (2013) 41–48.

     Received September 2024

CORRIGENDUM TO: QUANTUM ALGORITHMS FOR UNATE AND BINATE COVERING PROBLEMS WITH APPLICATION TO FINITE STATE MACHINE MINIMIZATION

ABDIRAHMAN ALASOW
Portland State University, Portland, OR, USA
alasow@pdx.edu

MAREK PERKOWSKI
Portland State University, Portland, OR, USA
mperkows@ee.pdx.edu

We, the authors of the paper entitled "Quantum algorithms for unate and binate covering problems with application to finite state machine minimization" published in Journal of Applied Logics in 2023, would like to clarify that Figure 13 in our paper use a variant of Grover diffusion circuit that is not a standard Grover diffusion operator for the Boolean oracles and the phase oracles of L.K. Grover as presented in [Grover 1996]. However, this variant of Grover diffusion circuit in this figure is the same as the quantum diffuser proposed by [ALI below], which is the so-called "Grover controlled-diffusion operator".

For this reason, we would like to publish this clarification as "Corrigendum" in the *Journal of Applied Logics.*

[ALI] A. A-Bayaty and M. Perkowski, "A concept of controlling Grover diffusion operator: A new approach to solve arbitrary Boolean-based problems," submitted to Scientific Reports, 2023. Nature.

.

Received

www.ingramcontent.com/pod-product-compliance
Lightning Source LLC
Chambersburg PA
CBHW080557090426
42735CB00016B/3266